AF883833

Key Challenges in Geography

EUROGEO Book Series

Series Editors

Kostis C. Koutsopoulos, European Association of Geographers, National Technical University of Athens, Pikermi, Greece

Rafael De Miguel González, University of Zaragoza & EUROGEO, Zaragoza, Spain

Daniela Schmeinck, Institut Didaktik des Sachunterrichts, University of Cologne, Köln, Germany

This book series addresses relevant topics in the wide field of geography, which connects the physical, human and technological sciences to enhance teaching, research, and decision making. Geography provides answers to how aspects of these sciences are interconnected and are forming spatial patterns and processes that have impact on global, regional and local issues and thus affect present and future generations. Moreover, by dealing with places, people and cultures, Geography explores international issues ranging from physical, urban and rural environments and their evolution, to climate, pollution, development and political economy.

Key Challenges in Geography is an initiative of the European Association of Geographers (EUROGEO), an organization dealing with examining geographical issues from a European perspective, representing European Geographers working in different professional activities and at all levels of education. EUROGEO's goal and the core part of its statutory activities is to make European Geography a worldwide reference and standard. The book series serves as a platform for members of EUROGEO as well as affiliated National Geographical Associations in Europe, but is equally open to contributions from non-members.

The book series addresses topics of contemporary relevance in the wide field of geography. It has a global scope and includes contributions from a wide range of theoretical and applied geographical disciplines.

Key Challenges in Geography aims to:

- present collections of chapters on topics that reflect the significance of Geography as a discipline;
- provide disciplinary and interdisciplinary titles related to geographical, environmental, cultural, economic, political, urban and technological research with a European dimension, but not exclusive;
- deliver thought-provoking contributions related to cross-disciplinary approaches and interconnected works that explore the complex interactions among geography, technology, politics, environment and human conditions;
- publish volumes tackling urgent topics to geographers and policy makers alike;
- publish comprehensive monographs, edited volumes and textbooks refereed by European and worldwide experts specialized in the subjects and themes of the books;
- provide a forum for geographers worldwide to communicate on all aspects of research and applications of geography, with a European dimension, but not exclusive.

All books/chapters will undergo a blind review process with a minimum of two reviewers.

An author/editor questionnaire, instructions for authors and a book proposal form can be obtained by contacting the Publisher.

The Key Challenges in Geography Series is now indexed in SCOPUS.

Danuta Piróg · Agnieszka Świętek
Editors

Future-Ready Geography

Goals, Problems and Challenges

euro
geo

Springer

Editors
Danuta Piróg
Department of Logistics, Geography of Services and Labour Market
Institute of Law, Economics and Administration
University of the National Education Commission
krakow, Poland

Agnieszka Świętek
Department of Logistics, Geography of Services and Labour Market
Institute of Law, Economics and Administration
University of the National Education Commission
Krakow, Poland

ISSN 2522-8420 ISSN 2522-8439 (electronic)
Key Challenges in Geography
ISBN 978-3-031-81267-5 ISBN 978-3-031-81268-2 (eBook)
https://doi.org/10.1007/978-3-031-81268-2

© The Editor(s) (if applicable) and The Author(s), under exclusive license to Springer Nature Switzerland AG 2025, corrected publication 2025

This work is subject to copyright. All rights are solely and exclusively licensed by the Publisher, whether the whole or part of the material is concerned, specifically the rights of translation, reprinting, reuse of illustrations, recitation, broadcasting, reproduction on microfilms or in any other physical way, and transmission or information storage and retrieval, electronic adaptation, computer software, or by similar or dissimilar methodology now known or hereafter developed.
The use of general descriptive names, registered names, trademarks, service marks, etc. in this publication does not imply, even in the absence of a specific statement, that such names are exempt from the relevant protective laws and regulations and therefore free for general use.
The publisher, the authors and the editors are safe to assume that the advice and information in this book are believed to be true and accurate at the date of publication. Neither the publisher nor the authors or the editors give a warranty, expressed or implied, with respect to the material contained herein or for any errors or omissions that may have been made. The publisher remains neutral with regard to jurisdictional claims in published maps and institutional affiliations.

This Springer imprint is published by the registered company Springer Nature Switzerland AG
The registered company address is: Gewerbestrasse 11, 6330 Cham, Switzerland

If disposing of this product, please recycle the paper.

Introduction

In today's world, continuous and radical changes impact society, the economy, and the natural environment—areas central to geographers' research. This evolving landscape opens new areas of inquiry and reinforces existing ones. Increasingly, discussions in geographical research pose the question: Are we "ready for the future"? The ongoing processes of globalization, digitization, integration, and disintegration shape and modify the drivers of human development, leading to multidirectional social changes. Growing global challenges such as climate change, food security, depletion of conventional energy sources, conflicts, wars, the refugee crisis, and pandemics have not only introduced new issues for geographical research but have also highlighted the necessity for geographers to revisit old topics under new conditions and with new research methodologies. Among these topics, the role of geographic education is paramount in preparing younger generations—Generations Z, Alpha, and beyond—to understand and responsibly manage the world.

The chapters in this volume address significant issues and outline key challenges facing geography as both a scientific discipline and a school subject.

The first half of the book (Chapters "Geography as a Compass for Understanding the World"–"Discussion Groups in Social Media as a Tool for Detecting Critical Educational Events: The Example of Polish Teachers of Geography") focuses on geographic education, addressing questions such as: What role will geography play in preparing society for the future? Which issues will become central in geographical education? How should geography, as both a scientific discipline and school subject, address global, regional, and local environmental and socio-economic problems to maximize its potential in applied sciences?

Chapter "Geography as a Compass for Understanding the World" introduces the research findings presented in the monograph. Piotrowska traces the evolution of geography through the centuries in terms of study areas and methodologies, underscoring its enduring importance as both a scientific discipline and a school subject. Piotrowska highlights geography's multidisciplinary nature, where space serves as the foundation for understanding and building knowledge about the world. This includes spatial orientation and exploration using advanced technologies like Geographic Information Systems (GIS). In social and environmental contexts, Piotrowska presents contemporary geography as a science that contributes to sustainable development, diversity, understanding of otherness, and respect for all cultures' achievements. The author critically examines modern geography curricula, assessing whether the knowledge and skills imparted in selected countries fulfil the ambitious goals set for school geography. In Chapter "Education in Socio-economic Geography in Poland as a Response to the Needs of the Labour Market", Kilar analyses the content of geographical education to evaluate how well it prepares students for future economic challenges. She compares the skills and knowledge students acquire during socio-economic geography classes with labour market expectations, questioning the practical utility of school geography in preparing students for future careers. Chapter "Looking Back to See What's Forward?—Heritage Education in Geography—New Understanding and Future of Students Training", authored by Świętek and Cybul, explores the promotion of cultural and natural heritage through geographical education as essential for preserving Earth's environmental resources and advancing human achievements. They highlight the importance of addressing generational differences in designing effective future geographical education, not only by updating content in response to global changes but also by considering the evolving profiles of students. The authors employ AI and semantic field methods to illustrate generational differences, using research on young people's perceptions of cultural heritage as an example.

The following three chapters focus on the practices and conditions shaping future educational processes. Scholten, Bienert and Mehren advocate for inquiry-based instruction in modern geography education, aligned with cognitive constructivism theory, currently regarded as the most effective. They critically assess teaching aids used to implement this theory in schools. Zisi and Klonari emphasize the importance of spatial thinking as a key geographic skill, advocating for its development from preschool age. Their empirical studies demonstrate the effectiveness of original teaching aids and pedagogical interventions aimed at enhancing preschoolers' spatial skills. However, even the best teaching methods and aids are ineffective without prepared and motivated teachers. In Chapter "Discussion Groups in Social Media as a Tool for Detecting Critical Educational Events: The Example of Polish Teachers of Geography", Bargieł examines "critical" moments in teachers' work, tracing their origins and scale to offer support for future challenges. Her analysis of social media content in private geography teacher discussion groups highlights pressing professional issues, including remote and hybrid teaching due to the COVID-19 pandemic, the war in Ukraine, increased numbers of non-native speaking students, and the lack of technical, didactic and psychological support for teachers under these conditions. She stresses that future-oriented educational policies must prioritize teachers' mental well-being.

The third part of the book focuses on sustainable development, a crucial area in geographical research. It begins with a chapter by Brede, Eberth and Meyer, who propose "post-growth" as an alternative to the traditional economic growth paradigm, advocating for welfare that transcends economic growth. Their multi-stage research, utilizing focus group methods, underscores the importance of engaging youth with "agents of change" and nurturing transformative skills in support of Education for

Sustainable Development (ESD). In the subsequent chapter, Cichoń and Baarova identify twelve key competencies essential for young people to embody sustainable development principles in adulthood. They analyse factors facilitating these competencies, finding strong positive correlations between creative thinking and curriculum delivery methods, particularly problem- and project-based learning. The section concludes with a chapter by Azis, offering a comprehensive review of academic literature and international regulations concerning sustainable development in urban contexts. The findings provide valuable insights for urban planning and development of city infrastructure, land use and spatial management in alignment with sustainable development principles.

The final part of the book highlights diverse emerging phenomena and research challenges geographers will face. Montebelli addresses the growing urbanization pressure on the environment, suggesting that local communities should possess geographical knowledge of their surroundings, including unconventional living spaces. This knowledge, combined with local area planning, territorial awareness and active citizenship, can foster harmonious coexistence between communities and the environment. In Chapter "Mapping the Future Landscape: Ghettoization Patterns of Roma Population in Northwestern Bulgaria", a Bulgarian research team examines the impact of cultural diversity and discrimination on spatial changes. Their research on the Roma population's ghettoization patterns, featuring dynamic landscape evolution visualizations, demonstrates geography's capacity to elucidate social processes and inform anti-marginalization policies. The final chapter by Kumer and Mackenzie calls for a revision of geographical regularities in the social sphere. Their research among the queer community challenges the narrative that people living in post-socialist countries, in particular those in rural areas, are inherently less open to non-heterosexual individuals. They emphasize the significance of the local context and the communities' evolving nature.

For geography to remain a significant scientific discipline and an important school subject, it must respond swiftly and effectively to contemporary challenges. We believe this book will inspire action to enhance the quality of geographical research and education.

We wish you an inspiring read!

Danuta Piróg
Agnieszka Świętek

Contents

Goals and Challenges in Future-Oriented Geography Education

Geography as a Compass for Understanding the World 3
Iwona Piotrowska

Education in Socio-economic Geography in Poland as a Response to the Needs of the Labour Market 23
Wioletta Kilar

Looking Back to See What's Forward?—Heritage Education in Geography—New Understanding and Future of Students Training 41
Agnieszka Świętek and Piotr Cybul

Practices of Future Learning Processes

How to Develop Spatial Abilities in Kindergarten 65
Christina Zisi and Aikaterini Klonari

How Can Scientific Literacy Be Enhanced in Geography Education? Two Ways Forward: Geoboxes and Digital Experiments ... 85
Nina Scholten, Natalie Bienert, and Rainer Mehren

Discussion Groups in Social Media as a Tool for Detecting Critical Educational Events: The Example of Polish Teachers of Geography 101
Kinga Bargieł

Sustainability as a Future Challenge

Post-growth as a Topic in Geography Education—Suggestions from Secondary School Students in Germany in the Context of Education for Sustainable Development 133
Lara Brede, Andreas Eberth, and Christiane Meyer

The Importance of the University Environment in Shaping Competences Relating to Sustainable Development Among Geography Students in Poland and Czechia ... 153
Małgorzata Cichoń and Barbara Baarová

Defining and Measuring the Level of Sustainability for Developing World, A Review in Urban Prospective ... 177
Asad Aziz

Living in Future World—Different People in Different Spaces

New Participatory Educational Spaces. Shared Urban Regeneration and the Role of Geography in Lifelong Learning of and About the Territory ... 195
Stefania Montebelli

Mapping the Future Landscape: Ghettoization Patterns of Roma Population in Northwestern Bulgaria ... 215
Nadezhda Ilieva, Kaloyan Tsvetkov, Boris Kazakov, and Dimitar Enikiev

Queer Lives: Navigating Identity, Acceptance, and Challenges in Rural Post-socialist Slovenia ... 237
Peter Kumer and Shé Mackenzie Hawke

Correction to: How Can Scientific Literacy Be Enhanced in Geography Education? Two Ways Forward: Geoboxes and Digital Experiments ... C1
Nina Scholten, Natalie Bienert, and Rainer Mehren

Author Index ... 255

Subject Index ... 257

About the Editors

Danuta Piróg is an Associate Professor of Socio-Economic Geography at the University of the National Education Commission in Kraków. Her academic pursuits focus on optimizing geography education across all levels, with a special emphasis on evaluation; university-to-work transition and the professional training of future geography and science teachers; the career paths of school teachers and academic lecturers; and methods and tools for aptitude testing. She has served as the principal investigator on both international and national research projects concerning the demand for university graduates in the labour market and innovative strategies for teacher training. She is an expert for the Ministry of Education in the field of geography textbooks and serves as the Scientific Secretary of the Committee for Geographical Sciences at the Polish Academy of Sciences. As a lecturer, she provides training and professional development for geography and science teachers and supervises Ph.D. candidates conducting research in geography education.

Agnieszka Świętek, Ph.D. is a socio-economic geographer at the University of the National Education Commission in Kraków. Her research interests include multiculturalism, geography education, and multicultural education. She conducts research in the fields of "geography of encounter" and "geography of silence", examining how young people engage with cultural diversity and the influencing factors, such as formal education and media coverage. In her educational research, aligned with inquiry-based learning, she explores modern teaching methods, skill development, critical analyses of core curricula and educational materials, and the evaluation of geography teaching processes and services. As a lecturer, she teaches future geography teachers and conducts courses to develop interpersonal and multicultural skills among students from various degree programmes. She heads the Laboratory for Geography and Environmental Education and has participated in numerous international projects focused on creating innovative educational strategies and materials.

Goals and Challenges in Future-Oriented Geography Education

Geography as a Compass for Understanding the World

Iwona Piotrowska

Abstract What if geography changed the world? The questions posed by geography concern all citizens, as natural, political and social phenomena cannot be excluded. Geography is a discipline that primarily studies empirical reality, and it is characterized by the fact that it is more concerned with the concretes of everyday life, the concrete geographical space, than other scientific disciplines. Geography is the discipline responsible for studying particular areas of the Earth's surface and providing knowledge about them, among other things in the form of descriptions and maps. It is supposed to satisfy human curiosity by answering questions about what and where things are, why certain phenomena and processes occur where they do and not elsewhere. Based on a study of the literature, the concept of geography as a synthesis of knowledge from different disciplines is still very evocative for 21st-century geographers (McDougall 2003; Pitman in Geoforum 36:2, 2005; Lau and Pasquini in Geoforum 39, 2008; Demeritt in Geoforum 40, 2009). In the past, geography with its discoveries served countries in their expansion, so now it can be useful in the work of building a world whose unity will be based on the preservation of diversity, on the understanding of difference and on respect for the achievements of other cultures (Wilczyński in Ideowe źródła i tożsamość geografii. Wydawnictwo Naukowe Uniwersytetu Pedagogicznego, 2011). From the study of landforms and climate in the 1970s, it has moved on to spatial planning, understanding the strategies of multinationals or the phenomena of spatial segregation. Geography seems to be useful for understanding the world by stimulating the desire to travel, enabling an investment in the complexity of today's world where civic engagement is necessary, a shift from lived space to perceived space and then to designed space using geoinformation technologies, especially geographic information systems (GIS). A survey of school and university teachers will provide answers as to whether geographical education is up to the task.

I. Piotrowska (✉)
Laboratory of Geography Didactics and Educational Research, Faculty of Geographical and Geological Sciences, Adam Mickiewicz University in Poznań, Bogumił Krygowski 10, 61-680 Poznań, Poland
e-mail: ipiotrow@amu.edu.pl

© The Author(s), under exclusive license to Springer Nature Switzerland AG 2025
D. Piróg and A. Świętek (eds.), *Future-Ready Geography*, Key Challenges in Geography, https://doi.org/10.1007/978-3-031-81268-2_1

Keywords Geographical space · Geography of thinking · Competence of the future

Introduction

Geography fulfills a fundamental human need by offering a comprehensive understanding of the world. It is widely regarded as one of the oldest sciences, with roots tracing back to the earliest stages of human intellectual development. Eratosthenes (276 BC–194 BC) is credited as the originator of the term, with his geographical insights chronicled in the works of Strabo (63 BC–c. 24 AD). According to Wilczyński (2011), geography has existed for as long as humans have roamed the Earth's surface. Human curiosity has historically propelled exploration of the unknown, resulting in the accumulation of geographical knowledge passed down through generations. This knowledge, derived from observation, explanation, and interpretation, forms the foundation for comprehending the natural environment and the world at large. Immanuel Kant succinctly captured the value of geographic knowledge by stating that "Nothing can so enlighten a sound human mind as precisely geography" (as cited in Bergsten 1988, p. 69, translated by W. Wilczyński; back translated). In modern times, geography remains a vital discipline. It navigates the challenges posed by civilization, technology, and dynamic environmental changes, which in turn trigger various effects and contribute to the global transformation.

Geography is a chorological science that emphasizes space as the foundation of everything, cognition and building knowledge about the world, in which occurring phenomena are always related to a specific region (Schatzki 1991; Soja 2010; Adamczewska et al. 2023). Geography stands poised to confront future challenges, but it necessitates keen observation, attentive listening, and thorough explanation of events unfolding in geographic spaces across various scales—from local to regional to global. The inquiries of geography are universal, encompassing natural, political, and social phenomena that affect all citizens. Unlike other scientific disciplines, geography focuses primarily on empirical reality, emphasizing the tangible aspects of everyday life and geographic spaces. It is tasked with the study of specific areas of the Earth's surface, offering insights through descriptions and maps. Geography seeks to satiate human curiosity by addressing questions about the identities and locations of objects, the reasons behind phenomena and processes occurring in specific places, and their absence elsewhere.

After reviewing the literature, it is evident that in the twenty-first century, geographers still find the concept of geography as a synthesis of knowledge from various disciplines highly compelling (McDougall 2003; Pitman, 2005; Lau and Pasquini 2008; Demeritt 2009). Nałkowski (1925, p. 27) notably recognized and articulated the significant value of geographic synthesis. His notion of "picturesque geography", which integrates science and art, was founded on certain epistemological principles. W. Nałkowski believed that geography, by training individuals to associate, discern similarities and differences, and establish connections between seemingly disparate phenomena, cultivates what he termed "geographical sense". This innate ability to perceive the world holistically, elevated by W. Nałkowski to the status of a sense, forms the cornerstone of geographical knowledge. Therefore, geography assumes a paramount role in both learning about the world and comprehending and explaining observed phenomena and processes.

It is imperative that geography education cultivates the skills encapsulated in the notion of geographic sense, as delineated in the Polish school core curriculum (Journal of Laws 2018, item 467):

- Proficiently locate objects and phenomena and assess them from diverse perspectives;
- Discern, observe, comprehend, and appreciate natural and socio-economic phenomena and processes occurring at local, regional, and global scales, recognizing their interrelationships;
- Elucidate the causal factors and consequences of the diversity evident in nature and human life and activity;
- Evaluate the suitability of the natural environment for human needs and anticipate the directions of changes, both positive and negative, resulting from human economic activities;
- Appreciate the beauty of nature and the inherent values therein, while recognizing the imperative to judiciously utilize and safeguard the Earth's natural resources;
- Make informed decisions regarding solutions to local, regional, and global challenges;
- Demonstrate autonomy in navigating the geographical environment and actively engage in addressing issues pertinent to one's own region.

The human sense of geography enables us to perceive the world as interconnected wholes, where landscapes represent the synthesis of all natural and cultural phenomena. While traditional geography once fueled countries' expansion, today it holds potential for fostering global unity based on diversity preservation, understanding differences, and respecting other cultures (Wilczyński 2011).

Since the 1970s, geographic research has shifted from studying landforms and climate to encompass spatial planning, understanding multinational company strategies, and phenomena like spatial segregation.

According to Antoine (2010) for physical geography the 1970s as a significant turning point. New perspectives emerged at the sources of fundamental changes that, around the 1970s, renewed the approach to geography and opened it to the social sciences. Dia (2022) states that these changes will have a significant impact on physical geography and its relationship to risk. Approaches to geography have changed over time. Some geographers emphasized classical geography with reference to zonality (Rognon 1967; George 1972). The issue of zonality of processes and phenomena aimed to determine the role of morphogenetic processes in the formation of relief occurring in various areas of the globe (Davis 1899; Tricart 1965 after Klimaszewski 1978; Tricart and Cailleux 1965). Some attempted to renew physical geography and, moreover, to restore the unity of geography as a whole (Pinchemel 1988).

An important place is also occupied by Bertrand (2002), who emphasized the natural side of the nature-society interface. Already in 1968, G. Bertrand placed the concepts of landscape and geosystem at the center of integrated physical geography. He treated the landscape and geosystem not only as natural, but also stated that they also included all the consequences of anthropogenic activities. Applicational approaches in relation to G. Bartrand's ideas were found in the works of other geographers, such as Kostrzewski (1986), applying the theory of geosystem functioning to contemporary morphogenetic environments. One of the most important elements of the geoecosystem is the relief of the land, which determines its internal structure and determines the circulation of matter and energy flow. According to Dia (2022) stated that G. Bertrand redefined physical geography as an open field, leaving largely room for anthropogenic aspects. The emergence and consolidation of risk geography is a direct consequence of the evolution of the discipline itself and its scientific orientations in the twentieth century. According to him, although risk geography is a young field, geographers have long been interested in natural phenomena that cause damage to societies. Starkel (2008) also referred in his geographical and geomorphological research on slope and fluvial processes transforming the relief to the 1970s, wondering whether the 1970s model is still valid.

Therefore, it is necessary to strongly emphasize the unique role that geography played and still plays today. Geography is now instrumental in comprehending the complexities of the modern world, facilitating transitions from lived space to perceived space and eventually to designed space using geoinformation technologies, particularly geographic information systems (GIS) (Source: https://clio-cr.clionautes.org/a-quoi-sert-la-geographie.html). Geoinformation technologies not only enhance our understanding of the world but also enable investment in civic engagement. They allow us to navigate the intricacies of contemporary society, transcending mere travel stimulation. With GIS, we can delve deeper into the complexities of our surroundings, contributing to informed decision-making and active participation in shaping our environment. Geoinformation technologies facilitate multidimensional analyses of geographical elements, demanding critical thinking skills such as argumentation, justification, evaluation, and valuation. This evolution signifies a shift in the role of geography education for young learners. Their education hinges on comprehending the spatial data collection, aggregation, and interpretation processes, along with utilizing them effectively in daily life (Piotrowska et al. 2023).

The adept utilization of geoinformation technologies not only enhances spatial thinking but also fosters geospatial thinking, as noted by Jo and Muñiz-Solari (2015), Metoyer et al. (2015) and Gaździcki et al. (2018). Furthermore, it aids in fortifying both natural and visual-spatial intelligence. Consequently, geographic education must evolve beyond solely reinforcing geographic thinking, encompassing (1) relational thinking regarding human-nature interactions, (2) spatial cognition and imagination, (3) problem-solving, and (4) critical analysis (Szkurłat et al. 2018; Szkurłat and Piotrowska 2018).

The rapid advancement of technology necessitates modern geography education to prioritize geospatial thinking. This entails merging geographic and spatial thinking to visualize phenomena and interpret them using elements like location, distance, change, and spatial relations (van der Schee et al. 2015). By integrating geoinformation technologies into geography education, it becomes feasible to cultivate spatially literate students, as highlighted by Donert et al. (2016). Heidegger (1974, p. 147, after Adamczewska et al. 2023) stated that there is a special and inseparable relationship between space and man, in which *space does not stand in front of man, but—with reciprocity—shapes him.*

Imagine if geography held the power to reshape the world! Through literature research, this study delves into the potential of geography to enact change in the world. The goal is to understand the impact of geography as both a science and an educational discipline. The study posits that geography is well-suited for such a role, given its emphasis on fostering key future skills. Independently of other disciplines, geography incorporates the development of these skills into its educational framework. These skills include critical thinking, creativity, research, self-actualization, information literacy, systems thinking, communication, and reflection (OECD 2022; https://pisa2022-maths.oecd.org). By honing these abilities and prioritizing quality education, geography can equip young individuals to confront real-world challenges effectively. Additionally, attention should be given to other essential skills like digital literacy, data management, time management, emotional intelligence, and flexibility.

For numerous years, geography and geographic education have adhered to the foundational principles of lifelong learning outlined by Delors (1998):

- *Learning to live together*: Individuals ought to broaden their understanding of diverse societies, histories, traditions, spiritualities, and foster tolerance, cooperation, and conflict resolution.
- *Learning to know:* Individuals should acquire the tools for knowledge acquisition, recognizing that effective teaching involves the capacity to concentrate, utilize existing knowledge, and engage in critical thinking. The ultimate aim of education is to cultivate the pleasure of comprehension, exploration, and possession of knowledge.
- *Learning to do*: It is essential to possess the ability to apply knowledge practically, foster cooperation and communication skills to enable individuals to navigate unforeseen circumstances, collaborate within teams, and creatively shape the future.
- *Learning to be*: The primary objective is the comprehensive development of the individual. Education must prioritize nurturing independent thinking, fostering critical judgment, and cultivating emotions and imagination. Without these, there is a risk of dehumanizing the world.

Only in this manner does geography possess the potential to enact change in the contemporary world. Present actions shape the future, emphasizing the importance of forecasting and reflecting on current endeavors. Nevertheless, the inherent uncertainty of the future remains intertwined with human activities, including education, as articulated by Albert Camus: "*School prepares children for life in a world that does not exist*". Conversely, Nałkowski (1901, p. 27) posited that "*Geography brings human thought out of the narrowness and one-sidedness of parochialism, opens wide comparative horizons. It must be of interest to anyone who can think scientifically, feel aesthetically and act practically in the wider arena*". In light of these perspectives, it becomes apparent that geography can serve as a compass guiding action and understanding in the modern world, encompassing both natural and socio-economic dimensions.

Geography assumes the role of a compass through its endeavors in research, problem-solving, and action across various domains. These include addressing challenges related to the natural environment, such as climate change, water scarcity, desertification, and population migration. Additionally, geography engages with issues encompassing social and economic diversity, political conflicts, and sustainable development, thereby guiding efforts toward informed action and resolution.

Development of Geography as a Science

Geography is a multidisciplinary science that delves into the Earth's surface, exploring its spatial variations in both natural and socio-economic contexts, and examining the intricate interplay between the environment and human activities. According to Leszczycki (1962), geography is concerned with the spatial differentiation of physical-geographical and socio-economic geographical structures, analyzing their interconnectedness. With its dual focus, geography scrutinizes both the natural environment and human endeavors, making it a multifaceted discipline.

Historically, geography has evolved from natural science, also known as natural history or cosmography, which flourished during the Renaissance era. This period witnessed significant advancements in understanding the natural world. Two main strands emerged from natural science: mathematical natural science (Latin *scientia*), which laid the foundation for various disciplines of natural (exact) sciences, and humanistic natural science, alternatively known as natural history or liberal arts (Latin *artes*). It wasn't until the seventeenth century that geography reclaimed its ancient name, marking a distinct field of study dedicated to exploring the Earth's surface and its complexities.

The demand for geographic knowledge surged during the modern era, particularly in the age of exploration, as rulers invested in expeditions to discover new lands and access their markets and resources (Wilczyński 2011). During the Renaissance, known as the era of great discoveries, geography emerged as one of the most practically applied fields of knowledge. It played a pivotal role in shaping the realities of the modern world. These achievements heightened geography's popularity, drawing widespread interest and recognition. In fact, Napoleon's establishment of the first chair of geography and history at the Sorbonne in 1808 underscores its significance during this period. Geography's raison d'être at that time lay in its practical utility for expanding the territorial possessions and spheres of influence of various European nations. This elevated status led to geography being hailed as the "Mother of the Sciences" in the seventeenth and eighteenth centuries, a sentiment echoed by notable figures such as John Locke (McDougall 2003, p. 220).

Following the independence of geography as a distinct science, the period spanning from 1870 to 1970 marked its zenith, often referred to as the golden age of geography. During this era, there was systematic and sustained development of individual branches and departments within the field. Major directions in geographic research emerged in the late 19th and early twentieth centuries, many of which continue to shape modern geographical inquiry. These directions paved the way for the development of specific geographic disciplines, fostering specialization and leading to the formulation of numerous scientific theories, such as those concerning relief development and glaciation.

In addition to academic advancements, organized centers of geographical thought flourished, including chairs of geography at universities and geographical societies. The high level of development and specialization in geography in various countries gave rise to distinct national schools of geography, such as the American, German, Russian, British, and French schools. Furthermore, international cooperation was fostered through initiatives such as international geographic congresses and collaborative research endeavors. The establishment of the International Geographical Union facilitated such cooperation, with joint research efforts, particularly in polar regions, contributing significantly to the advancement of global geographical knowledge in the latter half of the twentieth century (Wilczyński 2011). The foundations of modern geography were established in the mid-nineteenth century, largely attributed to the contributions of two prominent German scholars: Alexander von Humboldt (1769–1859) and Karl Ritter (1779–1859). Their work marked a pivotal shift in geography from being purely descriptive to a science focused on exploring and explaining patterns and regularities in the natural environment. Embracing the principle of cause and effect, modern geography seeks to understand the underlying mechanisms driving these phenomena, formulate conclusions, and derive generalizations based on empirical observations and scientific inquiry.

A. von Humboldt, often hailed as the father of physical geography, pioneered a synthesis of Earth's natural knowledge, integrating key findings from various scientific disciplines of the early nineteenth century alongside his own extensive geographical explorations across North and South America, Europe, and Asia. According to Wulf (2015), Humboldt's approach combined meticulous observations and measurements with a profound appreciation for analysis, emotions, and impressions. He emphasized the vital role of imagination in truly comprehending nature, asserting that "nature can be truly understood (…) only by using the imagination" (Wulf 2015, p. 414). Humboldt also formulated a fundamental principle of geography, encapsulated in the enduring maxim "Unity in diversity", which continues to resonate and is embodied in the European Union's slogan today (Wilczyński 2011). In contrast, Karl Ritter focused on exploring the intricate relationships and interdependencies between human societies and the natural environment. He contended that human endeavors are heavily influenced by, and in turn, have profound impacts on the natural world. Ritter advocated for a regional perspective on phenomena and championed comparative studies, describing geography in this manner as Earth science. Within this framework, Ritter embraced an anthropocentric approach, emphasizing the central role of human activities in shaping geographical landscapes and dynamics.

In the latter part of the nineteenth century, geography solidified its position as an academic discipline, witnessing the establishment of geographical faculties and departments in universities worldwide. Notably, among the Polish scholarly community, there existed a cadre of eminent figures who grasped the profound significance of geographic inquiry. Figures such as Jan Śniadecki, Stanisław Staszic, Wacław Nałkowski, Eugeniusz Romer, Stanisław Pawłowski, Wincenty Pol, and Ludomir Sawicki made notable contributions to the field. Sawicki (1932), for instance, envisioned geography as a repository of broad, practical knowledge applicable to everyday life. He envisioned it as a cornerstone of general education, a wellspring of life wisdom.

The classical era of geography's development encompasses the period during which the foundational ideas of modern geography were widely accepted as valid or served as a reference point for newly emerging concepts. Spanning until the mid-twentieth century in most countries, classical geography focused primarily on tangible phenomena observed within landscapes. These phenomena included both those shaped by human economic activity and those existing independently of human influence. Changes in landscapes were explained through the lens of everyday realities, highlighting the interplay between the field and economic realities, as well as environmental characteristics. In contrast, modern geography delves into the study of relationships, systems, and structures.

Classical geographers held the belief in the unity of all things and phenomena, which led to an emphasis on synthesis, or interdisciplinarity, in their approach. This is reflected in the key concepts of classical geography. In contrast, modern geography tends to favor an analytical approach. Beyond being a scientific discipline adhering to certain methodological standards, classical geography also served as a form of art aimed at shaping social consciousness (Wilczyński 2011). However, contemporary geography encompasses the entire spectrum of geographic sciences, each with its own distinct object of study. As a result, new research disciplines have emerged from the foundation laid by classical geography.

Śleszyński (2020) outlined the contemporary (early twenty-first century) subject-methodological classification of geographical sciences in Poland, reflecting its dual nature. Geography is categorized within both natural sciences, specifically as physical geography, and socio-economic sciences, under socio-economic geography. As a synthesizing science, geography addresses problems across these domains. It is characterized by a monothematic approach in terms of subject matter, aiming to formulate regularities based on the repetition of phenomena. Additionally, geography adopts an idiographic perspective concerning regions, providing spatial descriptions of phenomena.

Joliveau (2013) posed an intriguing statement and question: "Geography is everywhere, but where are the geographers?" This prompts us to recognize that geography encompasses the entirety of the space in which humans reside, examining its elements, objects, phenomena, and processes. Geography serves to study and elucidate these aspects, offering avenues for exploration and supplying the empirical data essential for their interpretation.

In the nineteenth century, various geographical societies were established: the French Geographical Society in 1821, the British Royal Geographical Society in 1830, the American Geographical Society in 1852, and the Polish Geographical Society in 1918. In recent years, these societies have engaged in discussions about the role of geography in societies, reflecting on their years of existence and seeking to outline future paths of development. These reflections have been accompanied by numerous conferences, debates, and seminars. For instance, the oldest French Geographical Society celebrated its 200th anniversary, prompting questions about the direction of geography, its role in modern societies, and its purpose. At its 2021 conference, the French Geographical Society explicitly posed the question "A quoi sert la géographie?" (What is the purpose of geography?) (Source: https://socgeo.com/2021/03/23/urgent-report-du-colloque-a-quoi-sert-la-geographie/). At the time, it emphasized the importance of discussing geography and promoting it as a dynamic discipline that is relevant to people's daily lives and "serves peace above all". In response to the question of what geography can contribute to the modern world, several proposals were put forward:

- Encouraging fieldwork from an early age, beginning in school;
- Utilizing maps to visually represent the natural and socio-economic environment;
- Modeling environmental features like settlements;
- Exploring the geopolitics of global warming, including examining responsibilities for its causes and effects;
- Investigating cyberspace and digital environments from a geographic perspective;
- Understanding the relationship between geography as a fundamental science and operational urbanism
- Addressing environmental protection concerns through geographical analysis and solutions.

Michon and Pitte (2021) proposed exploring geography through the lens of the five senses to restore equilibrium (…) in our perception of the world and reconnect with the classical heritage of geography, particularly in terms of landscape studies.

Polish geographers have engaged in ongoing discussions and research that bridge classical geography with contemporary environmental, social, and educational aspects. Noteworthy events include the 2018 Congress of Polish Geography in Warsaw, the 2022 Congress of Geographic Education in Łódź, and the 2023 Congress of Polish Geography in Poznań. These gatherings served as platforms for debating, synthesizing, and proposing new approaches, subsequently documented in numerous articles on geography as a science and geographic education (Figlus et al. 2022; Piróg 2022; Szkurłat et al. 2022a, b).

Chevalier (1997) conceptualized geographic knowledge as an open system, identifying distinct types of geography, or "poles," each catering to different social needs. These poles include university geography, applied geography, school geography, and public geography, collectively enriching, advancing, and disseminating scientific knowledge.

Primarily, the impact of geography as a scientific discipline, including its developed research methods, analyses, syntheses, and formulated laws, is most pronounced in applied geography. This influence extends to school geography, underscoring the undeniable connection between university and school knowledge levels. Adaptation and translation of scientific geographic content to educational levels are crucial, shaping the substantive and conceptual breadth. According to Chevalier (1997), "while university and school knowledge share a strong relationship, they are not mutually exclusive". Additionally, public perception of geography as both a science and a subject of study plays a pivotal role, as individuals should leverage scientific advancements effectively in their personal endeavors. This systemic perspective is further influenced by societal expectations and the interplay with other scientific disciplines. School geography operates with its unique logic and principles, distinct from university geography, as noted by Chevalier (1997, cited in Wastable 2010). However, Cabral (2024) states that school geography is an autonomous field of knowledge present in the school tradition and has a fundamental place in shaping Brazilian academic geography. In the functioning of the geographical environment, from a systemic perspective, the relationships and interactions between the natural (physical-ecological) and socio-economic systems are extremely important, which Bennett and Chorley (1978, after White et al. 1984) call systems interfacing, which also allows for understanding and explaining the contribution of human activity to the environment.

Geography as a Science and a School Subject

The twenty-first century recognizes the great potential of education in its broadest sense to educate a society and a knowledge-based economy. It is assumed that an important role in this process of educating society is played by geography, considered both as a scientific discipline and as a school subject, because by studying geographic space it makes it possible to explain the causes and consequences of its natural and socio-economic variation. An inspiring question was posed by Antoine de Saint-Exupéry in his publication "The Little Prince" (1943, 2000), while formulating an answer: "*And who is a geographer? A geographer is a scholar who knows the location of all the seas, rivers, towns, mountains, and deserts*". It is worth noting that geography is a discipline that goes beyond just naming and locating geographic objects. It encompasses investigation, analysis, explanation, forecasting, modeling, and education. Through geography, it is possible to develop the ability to perceive, evaluate and explain processes and phenomena occurring in the geographic environment, at a variety of spatial, temporal, and cultural scales. The unique value of this discipline is the holistic approach in the study of both the environment and human activities, and the perception of natural and cultural elements facilitates the explanation of changes occurring in the selected geographic space (Piotrowska 2010).

Cultural geography is a branch of geographic sciences that delves into the tangible and intangible aspects of space, regions, or landscapes, examining them from the human perspective as both creators and users. Its investigations center around cultural regions, cultural ecology, cultural integration, and cultural landscapes. Cultural geography explores the spatial dimension of culture, with a significant emphasis on analyzing symbols and signs, offering valuable insights into human interactions with their surroundings (Czepczyński 2007).

The rich cultural diversity found across the globe stems from a myriad of geographical, environmental, historical, and psychological factors. Geography identifies the notion of cultural circles, defined by geographic locations, which amalgamate specific cultural traits such as language, religion, material culture, and social and technological advancements. These circles evolve over time, often spanning centuries, shaping distinct worldviews and guiding values of their inhabitants. Understanding cultural circles enables comprehension of diverse perspectives and values held by different societies. Additionally, the concept of geographic thinking aids in navigating this complexity (Niesbett 2011; Piotrowska 2017). Educating geography at various educational levels becomes crucial in fostering awareness and appreciation of this diversity amidst global developments. Therefore, modern geography and schools face the imperative task of adequately educating, understanding, and sensitizing individuals to emerging cultural issues.

The study of culture by geographers sheds light on how individuals perceive, value, and utilize space according to their unique needs, as well as the factors influencing these needs. The distinctive development, character, and activities of people in cities and regions worldwide are intricately tied to culture, and cultural geographers are adept at elucidating this diversity. Thus, a cultural perspective provides a comprehensive understanding of processes rooted in human cognition and behavior, shaping our perception of a multifaceted world. Introducing these concepts into education early on increases the likelihood of nurturing young individuals who possess knowledge and understanding of the complexities of the modern world. For geography to fulfill its purpose and be a discipline useful in the modern world, it is crucial to develop geographical skills. Bański (2013) and Piróg (2022) wondered what contemporary and future geography, as well as geographical education, should be like, what to pay attention to, what needs to be discussed and noticed, and what challenges it faces (Piotrowska et al. 2019).

Geography's relevance in the modern world hinges on the development of geographic skills, underscoring the pivotal role of geographic education in human advancement. Skills, defined as "*the ability acquired through learning to perform tasks and solve problems pertinent to a particular field or profession*" (Law of December 22, 2015 on the Integrated Qualification System), are fundamental in this context. This definition aligns with the understanding of skills within qualification frameworks such as the European Qualification Framework (EQF) and the Polish Qualification Framework (PRK), where skills encompass the application of knowledge and know-how to tackle tasks and challenges effectively. Skills encompass both cognitive, involving logical, intuitive, and creative thinking, and practical, involving proficiency and utilization of methods, materials, tools, and instruments. In the context of Polish elementary education, basic geographic skills include:

- *Conducting field observations and measurements, analyzing the gathered data, and drawing conclusions based on the findings;*
- *Utilizing various tools such as plans, maps, photographs, drawings, charts, diagrams, statistical data, source texts, and information and communication technologies to gather, process, and present geographic information;*
- *Analyzing and interpreting maps depicting diverse content to discern geographical patterns and relationships;*
- *Identifying and understanding the connections and interdependencies among different aspects of the natural, socio-economic, and cultural environment;*
- *Formulating hypotheses and theorems to explain geographical regularities and phenomena;*
- *Making generalizations;*
- *Making informed evaluations of socio-cultural and economic phenomena and processes, both locally and globally;*
- *Generating questions, hypotheses, and solutions to problems concerning the geographic environment;*
- *Embracing new challenges and promoting environmentally and socially responsible actions;*
- *Developing spatial perception skills and spatial imagination for understanding geographical concepts;*
- *Engaging in constructive collaboration and enhancing communication abilities;*
- *Applying acquired geographical knowledge and skills in everyday life contexts;*

(Regulation of the Minister of National Education of February 14, 2017, Journal of Laws. 2017, item 356, pp. 116, 117).

In Polish secondary schools, the following are identified at the primary level:

- *Utilizing plans, physical-geographic and thematic maps, photographs, aerial and satellite images, drawings, charts, graphs, diagrams, statistical data, source texts, information and communication technology (ICT), and geoinformation technology (GIS) to gather, analyze, and present geographic information;*
- *Interpreting the content of various maps;*
- *Identifying relationships among different elements of the environment (natural, socio-economic, and cultural);*
- *Formulating hypotheses about the fundamental regularities of the geographic environment;*
- *Assessing political, socio-cultural, and economic phenomena and processes occurring in Poland and various regions of the world;*
- *Predicting the impacts of human economic activities on the geographic environment;*
- *Conducting mathematical calculations related to physical and socio-economic geography;*
- *Applying acquired geography knowledge and skills in everyday life while adhering to principles of sustainable development.*

(Regulation of the Minister of Education dated January 30, 2018, Journal of Laws, 2018, item 467, p. 163).

With reference to geographical cognitive and practical skills, Piotrowska and Abramowicz (2021) conducted an analysis of documents of selected European core curricula: English (National Curriculum in England, Primary Education, Geography programs of study: key stages 1 and 2. Department for Education, 2013; National Curriculum in England, Secondary Education, Geography programs of study: key stage 3. Department for Education, 2013), French (Le "Socle commun de connaissances, de compétences et de culture" constitue l'ensemble des connaissances, compétences, valeurs et attitudes nécessaires pour réussir sa scolarité, sa vie d'individu et de futur citoyen; Ministère de l'Éducation nationale, de la Jeunesse et des Sports; Le Bulletin officiel l'Éducation nationale, de la Jeunesse et des Sports) and Spanish (Currículo básico de la Educación Primaria, Real Decreto 126/2014, de 28 de febrero, por el que se establece el currículo básico de la Educación Primaria; Currículo básico de la Educación Secundaria Obligatoria y del Bachillerato Geografía e Historia, Real Decreto 1105/2014, de 26 de diciembre, por el que se establece el currículo básico de la Educación Secundaria Obligatoria y del Bachillerato).

They all emphasize the importance of geography as a science important from the point of view of educating a modern citizen who understands the geographical environment and the phenomena and processes occurring in it. The importance of these skills is confirmed by key competences, which, according to the Council of the European Union, constitute proven efficiency in the use of knowledge, skills and personal, social and/or methodological abilities in situations related to work or study and professional or personal development (EU Key Competences and Basic Skills 2018).

Interesting research on supporting students' synthetic geographical thinking using mind maps was conducted by Zhijie Huang et al. (2024). The term synthetic geographic thinking comes from the latest Chinese version of the "Geography Curriculum Standards" (General High School Edition 2017, compulsory education 2022) and is one of the four key priority competencies in high school geography teaching. This study found that the use of mind maps improved students' synthetic geographic thinking.

It is reasonable to posit that geography, through well-crafted skills, guides the evolution of today's youth and equips them to engage in perceptive understanding of the natural world. This includes the capacity for multifaceted analyses, interpretations, comparisons, and diagnoses, leveraging geoinformation tools.

Geographic Education for Sustainable Development

The skills outlined above, when cultivated through geographic education aligned with the principles of sustainable development, will be pivotal. Sustainable development hinges on a well-functioning education system, as it holds the mandate of preparing students with the knowledge, skills, and attitudes necessary to thrive in the future job market.

The concept of sustainable development emerged in the 1987 Report of the World Commission on Environment and Development, "*Our Common Future*", which was further elucidated in key documents adopted during the 1992 United Nations Conference on Environment and Development in Rio de Janeiro. These documents, including the Declaration on Environment and Development and Agenda 21 (Earth Summit Conference), lay the foundation for international legal frameworks that define strategies for sustainable development at various levels, from global to local. Poland has embraced all European agreements on sustainable development and has been collaborating with the UN Commission on Sustainable Development since 1993. Consequently, sustainable development can be understood as an evolution of the notion of progress, transitioning from technocratic economic growth (production, consumption, technological progress) to eco-development (development planned and implemented taking into account environmental opportunities and effects), and eventually to a multidisciplinary and humanitarian concept (Universal Declaration of Human Rights). This approach places human beings at the center, emphasizing their entitlement to a healthy and productive life in harmony with nature, the prosperity of the global community, intergenerational equity, and the fulfillment of individual potential, centered around human well-being, environmental harmony, intergenerational justice, and individual fulfillment (PWN Encyclopedia).

The Sustainable Development Goals (SDGs) hold significant importance as they serve as a blueprint for reshaping and improving the world in a sustainable manner, meeting the needs of the present generation while safeguarding the environment for future generations. These goals build upon the Millennium Goals and were further articulated in the document "Transforming Our World: Agenda for Sustainable Development—2030", adopted and ratified in 2015. Contained within this document are 17 Sustainable Development Goals, accompanied by 169 associated actions aimed at collective achievement by various stakeholders including national governments, international organizations, non-governmental organizations, the scientific and business sectors, as well as citizens. These goals are organized around five key areas: people, planet, prosperity, peace, and partnership (5xP). Serving as the foundation for the Agenda 2030, they address economic, social, and environmental aspects, identifying and prioritizing the most significant challenges of our times (Adamczewska et al. 2023).

Efforts to integrate sustainable development principles into education, particularly within the realm of geography, are linked to diverse educational strategies and methods. However, according to B. Wójtowicz (2010), this *integration process has been progressing too slowly relative to the urgent needs, often occurring in a disjointed and inconsistent manner* (Wójtowicz 2010, p. 175). Over the past decade, there has been a noticeable increase in the incorporation of sustainable development principles across various educational initiatives and economic strategies. However, despite this progress, Granados-Sánchez (2022), referring to the 2011 GUNi report "*Higher Education in the World 4*" on sustainability in higher education, highlights that the integration of sustainability topics into education has been relatively sluggish given the urgent global challenges humanity faces. This integration entails not only changes in teaching and learning methods but also shifts in management practices, community engagement, and knowledge creation and research approaches. As stated by Odell et al. (2020), educators and scholars should embrace a comprehensive and systemic perspective that fosters ethical transformation. This shift in educational approach would redefine methodologies and foster a culture of values and character development, transitioning from mere information transmission to participatory teaching and active learning. Such endeavors could culminate in the creation of a society-centric world—one that individuals aspire to inhabit.

The globally adopted Sustainable Development Goals (SDGs) hold immense significance for individuals worldwide and should be comprehended by people of all ages, including adults and youth. Research conducted by Leiva-Bronda et al. (2022) among Spanish students at the University of Valencia reveals that many students are cognizant of the SDGs and perceive them as relevant to their daily lives. The authors of the study emphasize the importance of establishing connections between the SDGs and everyday concerns to advance their effective implementation further. Research conducted by Puertas-Aguilar et al. (2022) and Ammoneit et al. (2022) underscores the imperative of education for sustainable development and the necessity for adequately preparing educators. It is crucial to align theoretical modules with teaching competencies focused on sustainable development. Emphasizing research and experiential learning can play a pivotal role in effectively integrating sustainability into geographic education (Adamczewska et al. 2023). Scientific and research skills are highlighted as fundamental learning outcomes in geographic education concepts, as noted by the researchers.

Conclusion

Geography, with its historical role in facilitating countries' expansion, now holds promise in shaping a world founded on diversity preservation, appreciation of differences, and mutual respect for diverse cultural accomplishments. With its rich tradition of synthesizing knowledge across various disciplines, both natural and humanistic, geography stands uniquely poised for this task. It encompasses research directions spanning systematic streams such as general, physical, and socio-economic geography, as well as chorological streams like regional geography. The interdisciplinary approach inherent in geography remains appealing to the emerging generation of 21st-century geographers. Geography was once envisioned as a discipline offering broad and practical knowledge applicable to daily life. It was perceived as unique, spanning the entire spectrum of earthly reality, transcending disciplinary boundaries and bridging the gap between the natural sciences and the humanities. This widespread appeal and omnipresence of geography stem from its expansive subject matter, serving as an ideal entry point for both natural and humanistic studies.

Contemporary geography comprises a system of sciences dedicated to unraveling the causes and consequences of the natural and socio-economic diversity within geographic space. Its research endeavors to elucidate the relationships and interdependencies among various elements of both animate and inanimate nature, as well as between these elements and human activities. The evolution from mere descriptions to comprehensive explanations, and subsequently to higher levels of scientific inquiry, underscores geography's growing applicability, particularly in the realm of planning. The discipline of geography is undergoing an epistemological transformation, marked by the emergence of a new scientific paradigm known as the geotechnological paradigm. This paradigm represents a novel approach to scientific inquiry made feasible not only by technological advancements but also by the vast proliferation of spatial data, including big data, mining data, and crowdsourced data, now accessible to the public (Kerski 2015). In this context, understanding the quality of data has become indispensable. The fusion of geography with geoinformation issues amplifies the potential of geography and renders it increasingly relevant to everyday life. Modern geography, enhanced by geoinformation tools and technologies, is poised to tackle global challenges and address pressing issues such as environmental transformations, climate change, and their consequential impacts on both the natural world and human societies. In light of these developments, geography serves as a guiding compass, illuminating the path toward action and comprehension of the contemporary world, encompassing both natural phenomena and socio-economic dynamics.

References

Adamczewska M, Angiel J, Dybska-Jakóbkiewicz I, Dzięcioł-Kurczoba B, Groenwald M, Hibszer A, Mularczyk M, Piotrowska I, Piróg D, Sadoń-Osowiecka T, Tracz M (2023) Edukacja geograficzna. Refleksje—wyzwania—badania. Wybrane aspekty. Prace Monograficzne Komisji Edukacji Geograficznej Polskiego Towarzystwa Geograficznego, Bogucki Wydawnictwo Naukowe 13:169

Ammoneit R, Turek A, Peter C (2022) Pre-service geography teachers' professional competencies in education for sustainable development. Educ Sci 12:42. https://doi.org/10.3390/educsci12010042

Antoine JM (2010) Géohistoire des catastrophes et des risques, une histoire del'environnement pyrénéen, mémoire de dossier d'habilitation à diriger des recherches. Université de Toulouse Le Mirail, Département de géographie et de l'aménagement

Bański J (2013) Jaka geografia?—uwarunkowania i spojrzenie w przyszłość. Przegląd Geograficzny 85(2):291–307

Bennett RJ, Chorley RJ (1978) Environmental systems: philosophy, analysis and control. Taylor & Francis, London, p 1978

Bergsten KE (1988) Geography: my inheritance. In: Hägerstrand T, Buttimer A (eds) Geographers of Norden, vol 52. Reflections on career experiences, Lund studies in geography, Ser. B, Human Geography, pp 61–70 (translated by W. Wilczyński)

Bertrand G (1968) Paysage et géographie physique globale. Esquisse Méthodologique, Revue Géographique des Pyrénées Et Du Sud-Ouest 39:249–272

Bertrand G (2002) Une géographie traversière. L'environnement à travers territoires ettemporalités, Paris, éditions Arguments
Cabral TM (2024) Geografia escolar brasileira e epistemologia da Geografia: teoria e prática na formação docente. Revista Brasileira de Educação em Geografia 14(24):05–34. Accessed 27 June 2024. https://doi.org/10.46789/edugeo.v14i24.1418
Chevalier JP (1997) Quatre pôles dans le champ de la géographie? Cybergeo: European Journal of Geography [En ligne], Epistémologie, Histoire de la Géographie, Didactique, Document 23. http://cybergeo.revues.org/6498. Accessed 2 June 2024
Czepczyński M (2007) Podejścia badawcze w nowej geografii kultury. In: Maik W, Rembowska K, Suliborski A (eds) Geografia a przemiany współczesnego świata. Wydawnictwo Uczelniana WSG
Davis WM (1899) The geographical cycle. Geol J 14:481–504
Delors J (red.) Edukacja—jest w niej ukryty skarb. Wyd. UNESCO; 1998
Demeritt D (2009) Geography and the promise of integrative environmental research. Geoforum 40
Dia S (2022) Rôle de la géographie physique dans l'émergence de la géographie du risque àl'Universite Cheikh Anta Diop de Dakar. Revue de géographie du laboratoire Leïdi. https://www.researchgate.net/publication/381317132_ROLE_DE_LA_GEOGRAPHIE_PHYSIQUE_DANS_L'EMERGENCE_DE_LA. Accessed 27 June 2024
De Saint-Exupéry A (2000) The little prince. Wyd. ALGO
Donert K, Desmidt F, Lázaro y Torres ML, De Miguel González R, Lindner-Fally M, Parkinson A, Prodan D, Wołoszyńska-Wiśniewska E, Zwartjes L (2016) The GI-learner approach: learning lines for geospatial thinking in secondary schools. GI_Forum 2:134–146
Figlus T, Musiaka Ł, Włodarczyk B, Wachecka-Kotkowska L (2022) Społeczna rola geografii jako nauki – wyniki badań empirycznych. Czasopismo Geograficzne 4:639–663. https://doi.org/10.12657/czageo-93-25
Gaździcki J, Gotlib D, Jażdżewska I, Zwoliński Z (2018) Aktualne aspekty edukacji geoprzestrzennej w Polsce. Rocznik Geomatyki 16, 3, 82:235–240
George P (1972) L'illusion quantitative en géographie, in La Pensée géographique française contemporaine. Mélanges offerts au professeur A. Meynier, Saint-Brieuc, Presses Universitaires de Bretagne, pp 121–131
Granados-Sánchez J (2022) Levels of transformation in sustainable curricula: the case of geography education. Sustainability 14:4481. https://doi.org/10.3390/su14084481
Heidegger M (1974) Budować, mieszkać, myśleć, Teksty: teoria literatury, krytyka, interpretacja. 6(18):137–152
Huang Z, Yang Y, Dong L, Gao A, Guo Z, Wang J, Zhang X, Zhao Y, Lu X (2024) Fostering students' geographical synthetic thinking using geographic subject mind maps. Humanit Soc Sci Commun 11(1). Accessed 28 June 2024. https://doi.org/10.1057/s41599-024-03312-x
Jo I, Muñiz-Solari O (2015) An agenda of GST in geography education for future. In: Muñiz-Solari O, Demirci A, van der Schee J (eds) Geospatial technologies and geography education in a changing world. Springer, Tokyo. https://doi.org/10.1007/978-4-431-55519-3
Joliveau T (2013) L'arbre de la géographie française. http://geotheque.org/wp-content/uploads/2013/03/arbre-de-la-g%C3%A9ographie-2015.png. Accessed 2 June 2024
Kerski JJ (2015) Opportunities and challenges in using geospatial technologies for education. In: Muñiz-Solari O, Demirci A, van der Schee J (eds) Geospatial technologies and geography education in a changing world. Springer. https://doi.org/10.1007/978-4-431-55519-3
Klimaszewski M (1978) Geomorfologia. PWN, Warszawa
Kostrzewski A (1986) Zastosowanie teorii funkcjonowania geosystemu do współczesnych środowisk morfogenetycznych obszarów nizinnych Polski Północno-Zachodniej. Spraw. PTPN 103:26–28
Lau L, Pasquini M (2008) Jack of all trades? The negotiation of interdisciplinarity within geography. Geoforum 39

Leiva-Brondo M, Lajara-Camilleri N, Vidal-Meló A, Atarés A, Lull C (2022) Spanish University students' awareness and perception of sustainable development goals and sustainability literacy. Sustainability 14:4552. Accessed 2 Dec 2023. https://doi.org/10.3390/su14084552
Leszczycki S (1962) Geografia jako nauka i wiedza stosowana, PWN
McDougall WA (2003) Why geography matters... but is so little learned. Orbis 47:2
Metoyer SK, Witham Bednarz S, Bednarz RS (2015) Spatial thinking in education: concepts, development, and assessment. In: Muñiz-Solari O, Demirci A, van der Schee J (eds) Geospatial technologies and geography education in a changing world. Springer, Tokyo. https://doi.org/10.1007/978-4-431-55519-3
Michon P, Pitte JR (dir.) (2021) A quoi sert la géographie? PUF 441. https://clio-cr.clionautes.org/a-quoi-sert-la-geographie.html. Accessed 2 Dec 2023
Nałkowski W (1901) Ziemia i człowiek. Szkice i studya geograficzne
Nałkowski W (1925) Zarys metodyki geografii. Wyd. M. Arcta
Niesbett RE (2011) Geografia myślenia. Dlaczego ludzie Wschodu i Zachodu myślą inaczej? Mistrzowie Psychologii, transl. E. Wojtych. Smak Słowa
Odell V, Molthan-Hill P, Sterling S (2020) Transformative education to address all sustainable development goals. In: Quality education, encyclopedia of the UN sustainable development goals. Springer
Pinchemel G (1988) et P. La face de la Terre, Paris A. Colin
Piotrowska I (2010) Rola dydaktyki geografii w kształceniu twórczego nauczyciela. In: Kwatera A, Cieśla P (eds) Rola i zadania dydaktyk przedmiotowych w kształceniu nauczycieli. Uniwersytet Pedagogiczny w Krakowie
Piotrowska I (2017) Geograficzne kształcenie w poznawaniu i rozumieniu zróżnicowania kulturowego na świecie (*kręgi kulturowe, geografia myślenia*). In: Rodzoś J, Szkurłat E (eds) Edukacja geograficzna wobec problemów współczesnego świata, vol 7. Prace Komisji Edukacji Geograficznej PTG Wydawnictwo Naukowe Bogucki, pp 39–51
Piotrowska I, Cichoń M, Abramowicz D, Sypniewski J (2019) Challenges in geography education—et review of research problems. Quaestiones Geographicae 38(1):71–84. Bogucki Wydawnictwo Naukowe, Poznań. https://doi.org/10.2478/quageo-2019-0009
Piotrowska I, Abramowicz D (2021) Czytanie i posługiwanie się mapą—kluczowa umiejętność w kształceniu geograficznym w podstawach programowych geografii na przykładzie Anglii, Francji, Hiszpanii i Polski. In: Hibszer A, Szkurłat E (eds) Kształtowanie i ocenianie umiejętności w edukacji geograficznej—założenia teoretyczne i praktyczna weryfikacja, vol 11. Prace Monograficzne Komisji Edukacji Geograficznej PTG, Bogucki Wydawnictwo Naukowe, pp 67–82
Piotrowska I, Przewoźna P, Bobińska A (eds) (2023) Technologie geoinformacyjne w edukacji geograficznej. Bogucki Wydawnictwo Naukowe
Piróg D (2022) Przyszłość i perspektywy rozwoju edukacji geograficznej w świetle studiów literatury. Czasopismo Geograficzne 4:583–606. https://doi.org/10.12657/czageo-93-23
Pitman AJ (2005) On the role of geography in earth system science. Geoforum 36:2
Puertas-Aguilar MA, Conway B, de Lázaro-Torres M-L, Rafael de Miguel González RM, Donert K, Lindner-Fally M, Parkinson A, Prodan D, Wilson S, Zwartjes L (2022) A teaching model to raise awareness of sustainability using geoinformation. Espacio, Tiempo y Forma, Serie 24 VI, Geografía 15:23–42. https://doi.org/10.5944/etfvi.15.2022.33687
Rognon P (1967) Le massif de l'Atakor et ses bordures (Sahara Central). Etudegéomorphologique. Thèse Doctorat d'Etat ès lettres. Editions du CNRS, in-4 broché, Paris
Sawicki L (1932) Zarys ogólnej geografii ziem polskich. Wyd. Orbis
Schatzki TR (1991) Spatial ontology and explanation, annals of the association of American geographers. 81(4):650–670
Śleszyński P (2020) Geografia wobec wyzwań systemowej reformy nauki w Polsce. Prace i studia geograficzne 65.2
Soja EW (2010) Seeing nature spatially. In: Albertson D, King C (eds) Without nature? A new condition for theology. Fordham University Press, New York, NY, pp 181–202

Starkel L (2008) Procesy stokowe i fluwialne przekształcające rzeźbę—czy model z lat 70. jest nadal aktualny? Land An 9:16–20
Szkurłat E, Piotrowska I (2018) GIS w nowej podstawie programowej geografii. Acta Universitatis Lodziensis, Folia Geographica Socio-Oeconomica, p 34
Szkurłat E, Hibszer A, Piotrowska I, Rachwał T (2018) Podstawa programowa z geografii źródłem nauczycielskich wyzwań. In: Hibszer A, Szkurłat E (eds) Nauczyciel geografii wobec wyzwań reformowanej szkoły. Prace Komisji Edukacji Geograficznej PTG 8
Szkurłat E, Adamczewska M, Dzięcioł-Kurczoba B (2022a) Postrzeganie edukacji geograficznej i jej społecznej roli w Polsce. Czasopismo Geograficzne 4:665–701. https://doi.org/10.12657/czageo-93-26
Szkurłat E, Hibszer A, Piotrowska I, Sadoń-Osowiecka T (2022b) Kształcenie nauczycieli geografii w Polsce w latach 2000–2022. Przyczyny i skutki kryzysu, propozycje rozwiązań. Czasopismo Geograficzne 4:607–638. https://doi.org/10.12657/czageo-93-24
Tricart J, Cailleux A (1965) Introduction à la géomorphologie climatique. Traité de géomorphologie. tome I. Ed. SEDES, Paris
Wastable M (2010) La notion de *belt* dans la géographie scolaire française des Etats-Unis. *Cybergeo: European Journal of Geography* [En ligne], Epistémologie, Histoire de la Géographie, Didactique. document 495. http://cybergeo.revues.org/23009. Accessed 20 Dec 2023
Wilczyński W (2011) Ideowe źródła i tożsamość geografii. Wydawnictwo Naukowe Uniwersytetu Pedagogicznego
White ID, Mottershead DN, Harrison SJ (1984) Environmental systems. Allen & Unwin, London
Wójtowicz B (2010) Geografia Rozwój Zrównoważony Edukacja Ekologiczna, Wydaw. Naukowe Uniwersytetu Pedagogicznego
Wulf A (2015) Człowiek, który zrozumiał naturę (transl. K. Bażyńska-Chojnacka, P. Chojnacki). Wydawnictwo Poznańskie sp. z o.o.
van der Schee J, Trimp H, Béneker T, Favier T (2015) Digital geography education in the twenty-first century: needs and opportunities. In: Muñiz-Solari O, Demirci A, van der Schee J (eds) Geospatial technologies and geography education in a changing world. Springer. https://doi.org/10.1007/978-4-431-55519-3

Legal Acts

Currículo básico de la Educación Primaria, Real Decreto 126/2014, de 28 de febrero, por el que se establece el currículo básico de la Educación Primaria.
Currículo básico de la Educación Secundaria Obligatoria y del Bachillerato Geografía e Historia, Real Decreto 1105/2014, de 26 de diciembre, por el que se establece el currículo básico de la Educación Secundaria Obligatoria y del Bachillerato.
EU Key Competences and Basic Skills, 2018. https://ec.europa.eu/education/policies/school/key-competences-and-basic-skills_en (dostęp: 20.06.2024)
Law of December 22, 2015 on the Integrated Qualification System
Le Bulletin official de l'Éducation nationale, de la Jeunesse et des Sports
https://www.education.gouv.fr/le-bulletin-officiel-de-l-education-nationale-de-la-jeunesse-et-des-sports-89558
https://www.education.gouv.fr/bo/20/Hebdo31/MENE2018714A.htm
Ministère de l'Éducation nationale, de la Jeunesse et des Sports
National Curriculum in England, Primary Education, Geography programmes of study: key stages 1 and 2. Department for Education, 2013
National Curriculum in England, Secondary Education, Geography programmes of study: key stage 3. Department for Education, 2013
Regulation of the Minister of National Education of February 14, 2017, Journal of Laws. 2017, item 356, pp 116, 117

Regulation of the Minister of Education dated January 30, 2018, Journal of Laws. 2018, item 467

Digital Sources

https://clio-cr.clionautes.org/a-quoi-sert-la-geographie.html
https://socgeo.com/2021/03/23/urgent-report-du-colloque-a-quoi-sert-la-geographie/
https://pisa2022-maths.oecd.org

Education in Socio-economic Geography in Poland as a Response to the Needs of the Labour Market

Wioletta Kilar

Abstract Education is one of the basic factors supporting socio-economic development and the development of the knowledge-based economy because it ensures the inflow of qualified employees to the labour market. Therefore, it is particularly important to select the education content and methods of its implementation. This is essential in order to effectively achieve the assumed goals and prepare young people to be able to meet the needs of the modern labour market, shaping the expected competences, e.g. creativity or critical thinking skills. The aim of this article is to present the content of education in the field of socio-economic geography in primary and secondary schools in Poland included in the applicable national core curriculum of the subject of geography and to assess the adequacy of this content to the expectations set by employers. I will strive to answer the following questions: what is the role of geographical education in preparing a young person to enter the labour market, but also to what extent is school education in the field of socio-economic geography adjusted to the employees' expectations towards graduates of schools functioning today and in the future. When preparing the article, the following research methods were used: query and analysis of the literature on the subject of the research, content analysis and analysis of the existing data (desk research).

Keywords Geographical education · Education in Poland · Competences · Labour market · Socio-economic geography

Introduction

Education is one of the key factors supporting socio-economic development but it is also the basis for the development of the knowledge-based economy as it ensures the inflow of qualified employees to the labour market. At the same time it should be noted that education is a necessary condition for the development of the information society. Following Toffler (1998: 221), we can assume that "the information society is focused

W. Kilar (✉)
University of the National Education Commission, Krakow, Poland

© The Author(s), under exclusive license to Springer Nature Switzerland AG 2025
D. Piróg and A. Świętek (eds.), *Future-Ready Geography*, Key Challenges in Geography, https://doi.org/10.1007/978-3-031-81268-2_2

on knowledge and competence as well as meeting the individual psychological needs of each individual". In his opinion, education should be future-oriented (Piontek and Buczek 2014).

The teaching process in school education takes place in Poland on the basis of national assumptions contained in the core curriculum document, which is published in the form of a resolution of the Polish Minister of National Education (Kolanowska 2020). The core curriculum includes both general assumptions included in the educational objectives—general requirements, as well as detailed provisions that can be found in the teaching content—detailed requirements as well as the conditions and methods of their implementation. Therefore, the basic and key element in formal education in Poland is the core curriculum.

The key goal of education from an early age should be to prepare a young person to function in the modern world, including entering the labour market in adulthood. In this context, it is particularly important to select the educational content and methods of its implementation in order to effectively achieve the assumed goals and shape the expected competences that are important on the labour market both now and those that will be important in the future, e.g. creativity or critical thinking skills.

The aim of the undertaken research is to diagnose to what extent the content of education in the field of socio-economic geography in primary and secondary schools, included in the applicable national core curriculum in geography, is adapted to the expectations set by employers on the labour market. I will aim to answer the questions: what is the role of geographical education in preparing a young person to enter the labour market, but also to what extent is school education in the field of socio-economic geography adjusted to the expectations towards graduates of schools operating today and in the future on the labour market?

To achieve the research goal, the following research methods were used: query and analysis of the literature on the subject of the research, content analysis and analysis of existing data (desk research).

Education for the Needs of the Labour Market in Literature

The conditions of functioning on the labour market are constantly changing and will undergo further changes. Therefore, an important issue is to prepare young people for functioning in conditions of the constantly changing environment.

The issue of the relationship between education and economic development as well as between education and the labour market is a research topic in many disciplines, but as research shows, "education is less and less able to keep up with the changes and needs generated by employers. (…) Moreover, there is great complexity and validity of the connections between education and the labour market" (Wawrzonek 2020: 260–261).

One of the greatest challenges of education is to develop common educational and market goals together with common strategies for their implementation.

Education as the Basis for Socio-economic Development

Man is the initiator as well as participant of the changes taking place in the economy. However, as Runde et al. (2023: 1) emphasize "education is the basis of any functioning society: it prepares people to live cohesively in a community, participate in political activities and make an economic contribution. Investments in education bring individual, national and global benefits. At the individual level, education can improve people's employability, earnings and health. Countries that have made significant investments in primary, secondary and higher education have been able to advance science and knowledge and create new products and technologies. Globally, investments in education the basis for social cohesion, economic growth, competitiveness and innovation. In times of rapid social, technological and economic change, investing in education is also important to build countries' resilience".

As research conducted since the 1960s shows, construed economic growth models assign an important role to human capital and in particular to the large role of education financed by the public sector (Arndt 1987). As Lubimow-Burzyńska (2016: 281) emphasizes, "education and human capital are considered key factors especially in developed countries, implementing the knowledge-based economy model". In building and functioning of a knowledge-based economy (KOW), the quality of human capital plays a key role, the basis of which is the education. The reason for this is that better educated people (work force) are more mobile and flexible, it is easier for them to carry out new tasks and acquire new skills, they are more creative, but they are also able to use a wider range of technologies (including newly developed ones) and specialized equipment. Therefore education is the basis for supporting economic development and also forms the foundation on which the economic and social well-being of citizens is built. Education is also the key to increasing economic efficiency and social cohesion increasing the value and productivity of the workforce and also contributes to the country's competitiveness on the global market through changes in technology and production methods (Lubimow-Burzyńska 2016).

As Schultz (1961: 1) points out no country has achieved sustained economic development without significant investment in education and human capital. For "although it is obvious that people acquire useful skills and knowledge, it is not obvious that these skills and knowledge are a form of capital, and that this capital is in large a part of the product of purposeful investments, that it has grown in Western societies at a much faster rate than the conventional (inhuman) capital and that its growth may be the most characteristic feature of the economic system".

The above is confirmed by research conducted by Wang and Liu (2016: 353–354) for 55 countries in the world, for the years 1960–2009, which indicate that "regardless of whether it is a developed or still developing country education, human capital and economic growth showed a significant positive correlation. Therefore, to support the development of education, authorities at all levels should increase the share of educational funds in their financial expenditures, optimize the structure of education expenditures and improve the efficiency of the use of funds. (...) The authors emphasize that high-quality education at primary and secondary levels is a prerequisite for high-quality higher education. When it comes to primary education, there are huge rural-urban disparities and regional disparities. Lack of funding, high dropout rates and teacher loss are common in some underdeveloped areas. Therefore, the government should increase investments in primary education so that every citizen can benefit from primary education on uniform, high quality and in effect reduce the rate of illiteracy and semi-illiteracy. Secondary education is the key link between primary education and higher education, which is also a stage that is often ignored. Therefore, government should take a clear responsibility in this process and gradually increase the proportion of financial education funds".

As B. Skubiak (2013) emphasizes, in the 21st-century economy, knowledge is the most valuable resource that determines economic development. Nowadays, the importance of natural resources and low-skilled labour in achieving economic development is decreasing, while the role of human capital is increasing. As the author points out, the experience of economically highly developed countries proves that economic successes, the high position of these countries on the world market and the competitiveness of their economies are increasingly dependent on the level of society's knowledge and the ability to create and use technical, economic, IT, etc., in economic processes.

Assuming that intensifying processes of civilizational changes will increasingly change the existing professional, social and cultural structures, it should be concluded that the society educated in the previous phases of socio-economic development will have to adapt to these new conditions of functioning and a significant part in this respect will be to popularize and increase access to education (Zioło 2020).

To sum the above up, as shown by the researchers conducted so far, education is the basis for the development of both individuals and for national economies at various scales of spatial systems. It should be noted that nowadays both access to education and the pursuit of acquiring high-quality education are crucial.

Features of the Modern Labour Market in Poland

The labour market is constantly changing due to changes in the socio-economic environment. "It is an economic mechanism that regulates the volume of labour supply (the potential that is being offered by job candidates) and labour demand (expressed by employers in the form of the amount of work necessary for the proper functioning of their organization)" (Plewka 2015: 40). Employee's problems resulting from the situation on the labour market have been an important direction of research on various levels for years (Piróg 2015a, b).

The labour market in Poland has been relatively stable since the 1990s (Adamowicz 2022). Poland still remains a country with one of the lowest unemployment rates in the European Union (Figs. 1 and 2). The unemployment rate, calculated in accordance with the definition adopted by Eurostat, amounted to 2.7% in December 2023 in Poland, in comparison with 5.9% in the European Union and

6.4% in the EURO zone. Therefore, Poland ranks second among EU countries in terms of the lowest unemployment.

According to the announcement of the Polish Ministry of Family, Labour and Social Policy as of 12 February 2024, the registered unemployment rate at the end of January 2024 was at the level of 5.4% and is lower than the unemployment rate from a year ago (i.e. at the end of January 2023) by 0.1 percentage point. A stable low unemployment rate has been maintained in Poland for almost a decade (Fig. 2).

An important feature of the labour market in Poland and in the world is its globalization, which is manifesting on several levels. On the one hand, the number of jobs directly or indirectly dependent on exports from abroad and nowadays more specifically from outside the European Union (EU), is growing. On the other hand, globalization creates jobs not only in exporting companies, but also in companies supplying them with goods and services. "In 2019, Polish exports directed outside the EU supported 2.3 million of job places in Poland. Another 761 thousand Polish employees worked in positions related to exports from other EU countries to the world. This means that 19% of jobs in Poland depend on exports towards outside of the EU" (The impact of globalisation on..). At the same time, globalization has

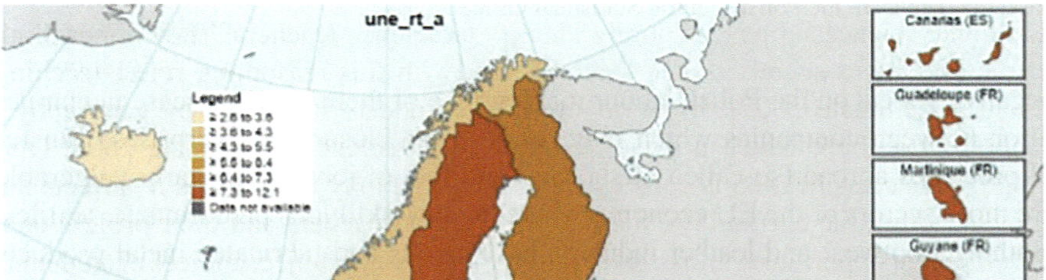

Fig. 1 Unemployment rate in European Union countries in 2023 (%). *Source* Eurostat (https://doi.org/https://doi.org/10.2908/UNE_RT_A)

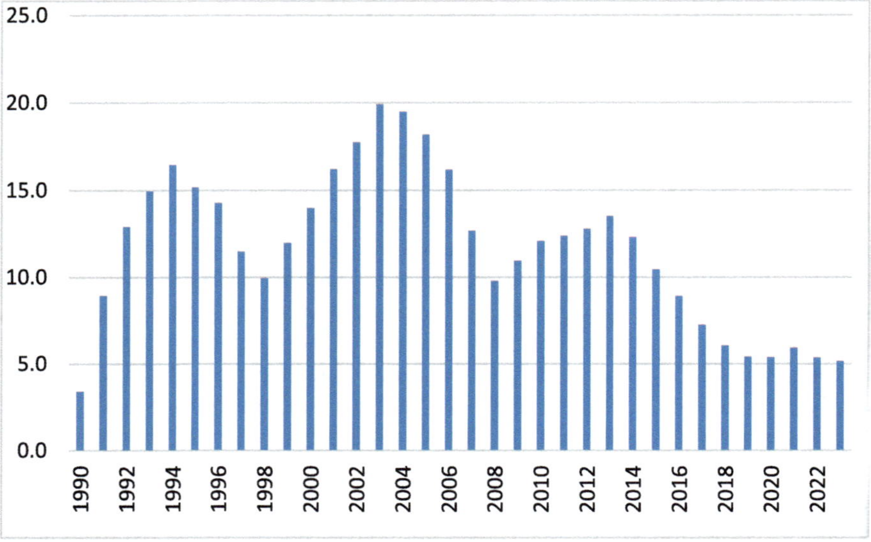

Fig. 2 Average registered unemployment rate in Poland in the years 1990–2023. *Source* Own study based on data from the Polish Central Statistical Office

negative effects on the Polish labour market. One of them is the increase in competition between companies which may result in the closure of enterprises, transfer of processes abroad (so-called offshoring) and loss of jobs. Particularly vulnerable are those sectors of the EU economy where the low-skill jobs predominate: textiles, clothing, footwear and leather industry, base metals and fabricated metal products and manufacturing. The manufacturing industry is most vulnerable to offshoring due to competition from low-wage countries. In the situation of progressing globalization the increase in company profits is no longer the result of increased labour productivity but is instead the result of cheaper labour costs (Księżyk 2005).

Moreover, one of the effects of the globalization process is the phenomenon of frequent job changes, as well as the replacement of an increasing number of employees with constantly developing intelligent technologies which can also be assessed in two ways (Wilsz 2014). We are also observing an increase in domestic outsourcing, which is reflected in the fact that a larger part of employees in OECD countries are legally employed by one company, but in practice work for another. Therefore, employees often physically work on the premises of one company, but their legal employer is an external company providing support services. Such employment relationships with a third party are often referred to as "domestic outsourcing", which is also a manifestation of the process of globalization of the labour market (OECD 2021).

Another important feature of the modern labour market is the phenomenon of "The Great Resignation", which is sometimes referred to as "The Great Rotation" by researchers of the Polish labour market (Kulisz 2022). This is a trend that migrated to Poland from the United States, where it was first observed. Great Resignation is a term promoted by organizational psychologist Professor Anthony Klotz (Texas A&M University), who observed it in the American labour market at the end of 2021, as a consequence of remote work dominating in many companies since 2020 due to the coronavirus pandemic. In the second half of 2021 it turned out that more than 20 million Americans left their jobs. Mass departures, or the Great Resignation, were to be a response primarily to poor employment conditions but also due to many other factors. In Poland, this phenomenon did not have such a massive scale among others due to the lack of sufficient funds that that would allow Poles to support themselves after resigning from work (Partyka 2023). In the situation of rising inflation and the war in Ukraine, employees focused on stability rather than on creating dream working conditions or offering of the tempting benefits. However, analysis of the labour market in Poland often refers to "The Great Rotation". This concept refers to the flow of employees to companies offering more secure and more prospective employment, with a certain employer who has features that are attractive to the employee at a given moment.

Another phenomenon also observed on the Polish labour market is the "quiet quitting", a trend on the labour market which, despite of its name, does not mean resigning from employment, but only performing a minimum of duties. It is associated to the lack of emotional attachment to the organization, treating work only as a source of earnings used for self-fulfilment in private life. This phenomenon is especially noticeable among young people just entering the labour market (Szczucka et al. 2023). Conducted research indicates that this trend is intensifying and there are many indications that it will remain an important challenge for employers in the coming years.

The "quiet quitting" trend results from, among others, the deteriorating mental condition of employees. According to the report prepared by Wellbee ICAN Institute in 2020–2023 on the mental condition of Polish employees, as many as 97% of employers have noticed a decline in employee motivation and engagement over the last 3 years. The worse the mental condition of employees, the lower is their effectiveness and commitment. A decrease in motivation and commitment often translates into an increase in their abstentions which negatively affects the work and results of employers (Report: Mental Condition.. 2023: 10). Therefore, the deterioration of employees' mental health also negatively affects their productivity, so it is necessary to involve employers in taking care of employees' mental health. According to research conducted by WHO (WHO Handbook … 2022), it is estimated that depression and anxiety of employees cost the global economy $1 trillion annually, mainly due to the lost productivity. "The research results clearly show the relationship between a high level of employees' mental well-being and the employer's profits. The return on investment (ROI) is estimated at 4:1. It is time for companies to realize that caring for the mental well-being of employees is an investment, not a cost" (Report: Mental Condition.. 2023: 29).

Research conducted by Jakubowski and Sitko-Dominik (2021) shows that the problem of poor mental health also concerns Polish teachers. This is the result of many factors, but is primarily due to the difficulty of working during the subsequent waves of the COVID-19 pandemic, during which teachers have experienced at least mild levels of stress, anxiety and depression, and due to remote working the distinction between professional and family life may have deteriorated (Jakubowski and Sitko-Dominik 2021; Świętek 2022).

To sum up, both the COVID-19 pandemic, the war in Ukraine, but also new trends in the global labour market have negatively affected employees on the Polish labour market, reducing their work motivation and commitment. Artificial intelligence (AI) is one of the key factors shaping the future of employment, but as history shows, each new technology usually creates more jobs than it eliminates. Therefore, it is necessary to look for new work models, greater flexibility in forms of employment and a balance between work and personal life.

Knowledge and Competences Expected on the Labour Market

Both the labour market and employers' requirements are constantly changing, which is the result of changing work conditions. Also the employers' approach to the knowledge, formal education and skills expected from employees are changing.

Currently, "even entering the labour market itself brings the need to modify and expand one's education acquired in the education system, not to mention the need to update the knowledge and improve skills in connection with the introduction of new technologies or the system of the production or organizational process. Hence, a natural need resulting from the realities of the modern labour market for employees is having appropriate learning skills" (Plewka 2015: 42).

According to The Future of Jobs Report from (2023), technology implementation will remain a key factor in business transformation over the next five years. The key skills employers expect from employees include analytical thinking and creative thinking. The next places on the list of sought-after skills are: resilience, flexibility and agility, and motivation and self-awareness; and curiosity and lifelong learning. Therefore, on the one hand, skills related to self-efficacy are valued, but also those related to the need of adaptation to disruptions in the workplace are also valued. Only next valued are the skills of dependability and attention to detail and technological literacy. The ten key skills also include those related to the attitude in working with others, such as empathy and active listening, as well as leadership and social influence.

As Przybyłka (2010) emphasizes, acquiring qualifications is the cheapest way to increase the level of competitiveness of the economy and enables accelerating the pace of economic development as well as the level of technological development and improvement of work safety.

Education becomes multidimensional and develops on the basis of four basic pillars:

- learning to know—learning for knowledge,
- learning to act—learning for work and competences,
- learning to live together—learning to understand others,

– learning to be—learning for your own development (Przybyłka 2010 after Zieliński 2007, p. 23).

All these dimensions are important and necessary, creating an inseparable whole that is influenced by today's dynamic labour market. However, the education system must adapt to changes taking place on the labour market and provide long-term education. However, education programmes both at the primary and secondary school levels and in higher education should be adapted to the requirements of the Polish labour market (Piróg 2012).

The need for lifelong learning is not only becoming a necessity driven by the needs of the labour market, but is also creating a need for changes in geographical education programmes (Matthews and Livingstone 1996). According to the recommendations of the Council of the European Union of 22 May 2018 on key competences for lifelong learning, there are eight key competences that citizens need for personal fulfilment, healthy and sustainable lifestyles, employability, active citizenship and social inclusion. It is therefore necessary to improve school education and to ensure excellent teaching quality, to further develop initial and continuing vocational education and training and to modernize higher education (Council of the European Union Recommendation 2018).

Education in the Field of Socio-economic Geography in the Light of the Core Curriculum. Correlation of Provisions in the Core Curriculum and Labour Market Needs

The contemporary, changing labour market in Poland and in the world, as shown by the presented research, values among others: creative and analytical thinking skills, employee flexibility and their ability to quickly adapt to new conditions (Christoph 2021). At the same time it brings many challenges related to the need for work-related mobility and the readiness to quickly move to new, different conditions, to another employer and another company. These are the challenges for which a large number of people operating on the labour market and preparing to enter the market are not prepared either formally or mentally. Therefore the entire education system faces a task that's goal should be to changed social awareness. It is particularly important to make people entering the labour market realize that when deciding on choosing a profession, they must be prepared for changes and for coping with the modern labour market, where working throughout life in one workplace or in one profession is not and will not be possible. Modern economy needs educated people who are able to work in conditions requiring quick retraining when necessary, as well as lifelong learning. Often the very act of entering the labour market brings in the need to modify and expand one's education acquired in the education system, but also to update knowledge and improve skills in connection with the introduction of new technologies or the modification of the production or organizational process. Hence, the key challenge resulting from the realities of the modern labour market is having learning skills (Plewka 2015).

Due to the fact that geographical education is complex in its nature and prepares one to understand the natural, but also socio-economic and cultural processes taking place on Earth but also how to function in a broadly understood geographical space, integrating other categories of space, which includes, among others, labour market, should be of key importance in the education system and thus in introducing a person to the labour market.

To achieve the assumed research goal, an analysis of the applicable provisions of the core curriculum for general education for primary schools (2017) and the core curriculum for secondary schools (2018) was carried out. In the Polish education system the core curriculum is the most important document, applicable throughout Poland on the basis of the regulation of the Minister of National Education, on the basis of which programmes of teaching, textbooks and the teaching frameworks are designed.

The core curriculum contains the general foundation for education in Poland, but also for individual school subjects, educational goals—general requirements, teaching content—specific requirements, as well as the conditions and methods of their implementation.

Geographical Education for the Needs of the Labour Market in Primary School

The analysis of the provisions contained in the current core curriculum in the Regulation of the Minister of National Education as of 14 February 2017 for primary schools in the field of geography, which is implemented in grades five to eight, indicates a significant predominance of detailed requirements focused primarily on acquiring knowledge and to a lesser extent skills of geographical abilities.

In the core curriculum:

(a) for educational purposes—general requirements include phrases such as: "identifying interdependencies between…", "determining regularities", "using source texts and information and communication technologies", "determining relationships and dependencies between", "evaluating phenomena and processes", but also "asking questions, formulating hypotheses and proposing solutions to problems…" Therefore, relatively high general expectations are placed on teenagers, but little attention is paid to how to achieve these goals. Developing these skills will certainly be helpful on the labour market because they closely relate to the analytical skills expected by the labour market, resulting in ability to solve complex problems, but also to use technology and reasoning. It seems crucial to use appropriate educational methods and to devote enough attention and time to developing the indicated skills to actually achieve this goal. At the same time, it is important that students not only acquire detailed knowledge, but also deepen it in individual thinking processes.

(b) for educational purposes—general requirements and terms are included that are directly related to entering and functioning on the labour market, such as:

- "undertaking constructive cooperation and developing communication skills with others",
- "recognizing one's predispositions and talents and developing passions and geographical interests",
- "overcoming stereotypes and shaping attitudes of respect, understanding and acceptance for other cultures".

(c) the teaching content—detailed requirements are dominated by educational goals that describe in detail what is the geographical knowledge that the student should acquire as a result of education, which most often begins with the verbs: characterizes, presents, understands, demonstrates, gives, indicates, distinguishes, lists, describes, determines. However, there are also provisions indicating the need to develop certain general skills that will also be needed at work, regardless of what additional specialized knowledge the employer will expect in the future. There are phrases such as: "determines dependencies...", "explains causes and consequences...", "assesses...", "determines similarities and differences...", "analyses changes...", "compares the size of...", "shows dependencies between ...", "identifies relationships".

Although they concern detailed geographical knowledge, it seems to be a great advantage that they are included in at least a few of the basic provisions and can also be a reference to other non-geographic spheres of life.

(d) in the teaching content—detailed requirements, it is also possible to find a few terms that indicate the need to acquire skills and develop attitudes needed in the modern labour market, e.g. "adopts an attitude of respect and understanding of other cultures".

To sum the above up, the natural provisions in the core curriculum for primary schools in the field of geography focus on detailed geographical knowledge and skills that a student should acquire in the education process. However, some educational goals and detailed requirements have been formulated in the core curriculum in such a way that students during geography lessons acquire not only knowledge and skills, but above all additional competences, which, when developed, will be helpful to them in subsequent stages of life, including on the labour market.

The requirements for students included in the analyzed core curriculum seem to be relatively high, considering that this subject in primary school is only available in the basic version, unlike in secondary school, where the student can also study extended geography. Therefore, in the detailed requirements, the wordings seem to require deep knowledge and are sometimes even abstract in nature, e.g. "assesses the socio-economic and cultural consequences of migration in Europe" or "is aware of the fact that it may have an impact on social, economic and cultural development of Poland in the future".

The provisions in the core curriculum for primary schools in the field of geography provide little evidence of creating the effect of acquiring knowledge and developing skills and attitudes that are expected in the modern labour market. Although a primary school student finishes this stage of education at the age of 15, he or she still has a few years to acquire certain skills and knowledge, I believe that shaping attitudes deeply embedded in the education system should start from an early age.

In the author's opinion, the biggest shortcoming of the current core curriculum in geography for primary schools is that it does not indicate the need to develop creativity in the education process. Among dozens of detailed requirements, only individual students and teachers are obliged to do so, e.g. by writing "designs of the tour route…". Nowadays, when access to knowledge is easy thanks to the high availability of technology, and it is devalued relatively quickly, and artificial intelligence is able to replace knowledge-based processes, shaping, deepening and developing creativity and critical analytical thinking should be key competences developed during all lessons and all school subjects from an early age. Unfortunately, too much focus on knowledge requirements means that the student and teacher do not have formal and time freedom in developing them and delving into selected issues, although this is in fact suggested in the conditions of implementation of the core curriculum.

Geographical Education for the Needs of the Labour Market at Secondary School Level

The applicable core curriculum for the third stage of education in geography is included in the Regulation of the Minister of National Education as of 30 January 2018, on the new core curriculum for general education for four-year general secondary schools, five-year technical secondary schools and two-year second-cycle vocational schools. This core curriculum for secondary schools has been in force since the 2019/2020 school year (for first grades at that time).

The initial provisions of the core curriculum included the objectives of general education and the most important skills that should be acquired by the student during general education. These include:

- thinking,
- reading,
- ability to communicate in your native language and foreign languages,
- creative problem solving,
- ability to efficiently use modern information and communication technologies,
- independent accessing of information, selection of information, synthesis and evaluation of information, reliable use of sources;
- acquiring habits of systematic learning, organizing acquired knowledge and deepening it;
- ability to cooperate in a group and ability to take individual actions.

Therefore, in the assumptions of education at the post-primary school stage, we find reference to the need for competences that are expected on the modern labour market.

However, the part of the core curriculum regarding geographical education is divided into basic and extended scope.

According to the analysis of the core curriculum for basic geography education at the secondary school level (secondary school and technical secondary school), the entries are divided into:

(a) educational objectives—general requirements which include primarily general formulations regarding geographical knowledge, skills and application of knowledge in practice, as well as shaping attitudes.

This part of the core curriculum, similarly to the provisions for primary school, lists the requirements that refer to the development of competencies expected on the labour market. Therefore helpful skills included in the educational objectives may be: formulating statements, assessing phenomena and processes, predicting effects, but above all, "critical, responsible assessment of changes…", "developing the ability to communicate and undertake constructive cooperation in a group".

Among the attitudes being shaped in geographical education, the following refers to the needs of the labour market: "…, awakening curiosity about the world", "shaping the attitude of understanding and respect…", "overcoming stereotypes and shaping the attitudes of solidarity, respect and empathy…".

(b) teaching content—detailed requirements, which mainly contain formulations focusing on activities necessary to acquire detailed geographical knowledge, and to a small extent, also skills.

While analyzing the detailed requirements, it seems that some of them that are being developed during geographical education may also be useful on the labour market, e.g. analyzing and explaining mechanisms, identifying factors influencing phenomena, explaining relationships, discussing the topic and analyzing spatial diversity.

However, only four of the detailed requirements are directly related to the labour market, but only indirectly to the actual development of competencies expected today on the labour market:

- characterizes the main features of the knowledge-based economy and factors influencing its innovation and development in Poland and other countries around the world;
- demonstrates the importance of human capital in economic development;
- discusses the manifestations and effects of the formation of the information society;
- explains changes in the employment structure, gives the causes of unemployment and analyzes the spatial differentiation of the labour market in Poland.

The analysis carried out indicates that compulsory geographical education at the secondary school level (both secondary school and technical secondary school) includes in the provisions of the core curriculum the development of competences necessary on the labour market in a very limited way. Indirectly, by acquiring geographical knowledge and skills, students will acquire competences such as: analytical skills, as well as develop skills in the field of new technologies. It is somewhat disturbing that only the general educational goals refer to the competencies expected by the labour market and there is no reference to them in the detailed provisions that teachers are obliged to implement during lessons. This may mean that in the process of geographical education at post-primary level, the teacher, as indicated in the basic provisions, will focus primarily on expanding geographical knowledge. However, due to time constraints and the lack of clear detailed requirements in the core curriculum, as indicated by the observation of the formal education process, he or she will not develop competences that are crucial on the labour market.

Conclusions

While analyzing the provisions of the core curriculum for primary and secondary schools in the field of compulsory geographical education, with particular emphasis on socio-economic geography in Poland, the author assessed the adaptation of the provisions in these basic documents in the Polish education system to the needs related to the need of developing the specific competences expected today by the labour market.

The presented diagnosis indicates that the educational goals and teaching content in the field of socio-economic geography and geography as a whole, both in primary and secondary schools, do not reflect the needs of today's labour market.

Therefore, it should be stated that compulsory geographical education in Poland at present prepares a young person to enter the labour market to a limited extent. In the face of rapid changes in the labour market, but also in the changing environment (development of technology and artificial intelligence, as well as widespread access to them in Poland and in other developed countries), it seems necessary to look at the education process from a slightly different perspective, which should not focus on only the existing models, mainly students acquiring detailed knowledge, but above all on acquiring key competences. It is obvious that without knowledge it is not possible to develop skills and broadly understood competences, but acquiring knowledge should not be the most important goal in itself.

Therefore, it seems necessary to give teachers more time and formal freedom in shaping the education programme in a given school and adapting it to the individual needs of a given group of students, so that they can devote more time to the issues they choose to and analyze them more deeply.

Moreover, in the provisions of detailed requirements, the perception of the above, should be changed in the context of detailed provisions focusing on acquiring knowledge. It will seem necessary to focus on developing specific competences with which a graduate should exit the compulsory education system and be able to enter the labour market. Therefore, more attention should be focused on shaping:

creative thinking, critical analysis skills, resistance to failure, flexibility, motivation, developing curiosity and the need for lifelong learning.

Nevertheless, changes in the education system take time and their effects may be visible only after a certain period. However, for deeper changes to occur, it is necessary to:

- ensure teacher education adapted to contemporary needs and conditions, which should take place on various levels and should be conducted according to a carefully planned programme, by people prepared for this goal, so that teachers not only acquire the knowledge needed for teaching, but also develop interpersonal, organizational and pedagogical competences,
- provide constant and systematic increase in expenditure on education,
- implement educational reform preceded by in-depth research and diagnosis of the current education system and necessary changes at various levels, as well as the introduction of a core curriculum based on the development of specific competences,
- provide schools with access to the necessary resources so that they can develop depending on their own needs (access to equipment, technology, all teaching aids).

Introducing changes in the perception of education seems necessary so that new graduates entering the labour market are prepared for the challenges they will face. The closed relationship between education and the labour market should apply not only to schools educating for specific professions, because ultimately all graduates will find themselves on the labour market. Therefore, regardless of the educational profile, the student should develop clearly defined competences.

We can ask ourselves a question: if the development of competences needed on the labour market does not take place during compulsory geographic education, is formal education putting any emphasis on the competencies expected on the labour market at all? Polish education system still focuses primarily on acquiring detailed knowledge and certain skills, but not on what a young person will need in adult life.

References

Adamowicz M (2022) Czynniki przemian rynku pracy w Polsce w latach 1989–2021. Rynek pracy wobec wyzwań przyszłości: ewolucja i współczesne uwarunkowania, pp 113–146

Announcement of the Polish Ministry of Family, Labour and Social Policy as of February 12, 2024 (Komunikat Ministerstwo Rodziny, Pracy i Polityki Społecznej z dnia 12 lutego 2024 roku nt. bezrobocia rejestrowanego w styczniu 2024 roku) https://www.gov.pl/web/rodzina/bezrobocie-rejestrowane-w-tyczniu-2024-roku--wstepne-dane. Last access 29 Mar 2024

Arndt H (1987) Economic development: the history of an idea, vol 3. University of Chicago Press, Chicago, pp 49–88.

Christoph M (2021) Zmieniająca się szkoła wobec wymagań rynku pracy, i nie tylko. Forum Pedagogiczne 11(1):159–177

Council of the European Union Recommendation of 22 May 2018 on key competences for lifelong learning (2018/C 189/01) https://eur-lex.europa.eu/legal-content/EN/TXT/HTML/?uri=CELEX:32018H0604(01)#d1e32-7-1. Last access 18 June 2024

Jakubowski TD, Sitko-Dominik MM (2021) Teachers' mental health during the first two waves of the COVID-19 pandemic in Poland. PLoS ONE 16(9):e0257252. https://doi.org/10.1371/journal.pone.0257252.PMID:34555049;PMCID:PMC8460021
Kolanowska E (2020) The system of education in Poland. Eurydice. Foundation for the development of the education system Polish National Agency for the Erasmus+ Programme
Księżyk M (2005) Globalizacja a rynek pracy z uwzględnieniem Polski. Nierówności społeczne a wzrost gospodarczy. 6:287–302 (https://www.ur.edu.pl/files/ur/import/Zeszyty/zeszyt-6/23_Ksiezyk_Marianna.pdf)
Kulisz M (2022) Great resignation–geneza, skutki i polski rynek pracy. https://wynagrodzenia.pl/artykul/great-resignation-geneza-skutki-i-polski-rynek-pracy. Last access 29 Mar 2024
Lubimow-Burzyńska I (2016) Znaczenie edukacji dla wzrostu gospodarczego—przegląd badań. Prace Naukowe Uniwersytetu Ekonomicznego We Wrocławiu 437:280–287. https://doi.org/10.15611/pn.2016.437.27
Matthews H, Livingstone I (1996) Editorial: geography and lifelong learning. J Geogr High Educ 20(1):5–9. https://doi.org/10.1080/03098269608709339
OECD (2021) The rise of domestic outsourcing and its implications for low-pay occupations. OECD employment outlook 2021: navigating the COVID-19 crisis and recovery. OECD Publishing, Paris. https://doi.org/10.1787/937ad5bc-en
Partyka M (2023) Great Resignation w Polsce—na czym polega Wielka Rezygnacja i jakie są jej skutki? https://betterworkplace.pl/blog/great-resignation-w-polsce-na-czym-polega-wielka-rezygnacja-i-jakie-sa-jej-skutki/. Last access 1 April 2024
Piontek B, Buczek P (2014) Integration of knowledge and education and its importance for management of socioeconomic development (Integracja wiedzy i edukacji i jej znaczenie dla zarządzania rozwojem społeczno-gospodarczym). Nierówności Społeczne a Wzrost Gospodarczy. 39(3):190–215
Piróg D (2012) Changes to the conception of geography curricula with-in university education in Poland in the 21st century in the face of labour market challenges. Eur J Geogr 3(2):24–41
Piróg D (2015a) The issue of labour research in the field of socio-economic geography in Poland and in the Anglophone countries. Stud Ind Geogr Comm Pol Geogr Soc 29(4):127–143. https://doi.org/10.24917/20801653.294.9
Piróg D (2015b) Przechodzenie absolwentów studiów geograficznych na rynek pracy: proces, determinanty, predykcja. Wydawnictwo Naukowe Uniwersytetu Pedagogicznego, Kraków
Plewka C (2015) Sytuacja ludzi młodych na współczesnym rynku pracy. Problemy Profesjologii 1:39–51
Przybyłka A (2010) System edukacji wobec potrzeb rynku pracy. Prace Naukowe Uniwersytetu Ekonomicznego we Wrocławiu. Rola państwa w gospodarce rynkowej na progu XXI wieku. 102:76–87
REPORT: Mental condition of Polish workers 2023 (RAPORT: Kondycja psychiczna polskich pracowników 2023). https://storage.googleapis.com/wellbee-ghost/reports/employee-mental-health/RAPORT_PL.pdf. Last access 1 April 2024
Regulation of the Minister of National Education as of February 14, 2017 for primary schools on the core curriculum for pre-school education and the core curriculum for general education for primary school, including for pupils with moderate or severe intellectual disabilities, general education for an upper secondary school, general education for a special needs school and general education for a post-secondary school (Rozporządzenie Ministra Edukacji Narodowej z dnia 14 lutego 2017 r. w sprawie podstawy programowej wychowania przedszkolnego oraz pod-stawy programowej kształcenia ogólnego dla szkoły podstawowej, w tym dla uczniów z niepełnosprawnością intelektualną w stopniu umiarkowanym lub znacznym, kształcenia ogólnego dla branżowej szkoły I stopnia, kształcenia ogólnego dla szkoły specjalnej przysposabiającej do pracy oraz kształcenia ogólnego dla szkoły policealnej) https://isap.sejm.gov.pl/isap.nsf/DocDetails.xsp?id=wdu20170000356. Last access 1 April 2024
Regulation of the Minister of National Education of 30 January 2018 on the core curriculum of general education for general secondary school, technical secondary school and upper secondary

school of the industry (Rozporządzenie Ministra Edukacji Narodowej z dnia 30 stycznia 2018 r. w sprawie podstawy programowej kształcenia ogólnego dla liceum ogólnokształcącego, technikum oraz branżowej szkoły II stopnia) https://isap.sejm.gov.pl/isap.nsf/DocDetails.xsp?id=WDU20180000467. Last access 1 Apr 2024

Runde DF, Bandury R, McLean M (2023) Investing in quality education for economic development, peace, and stability. Center for Strategic and International Studies

Schultz TW (1961) Investment in human capital. Am Econ Rev 51(1):1–17

Skubiak B (2013) Edukacja jako czynnik wspierający rozwój gospodarczy. Implikacje dla Polski. Studia Ekonomiczne 129:195–203

Szczucka A, Jelonek M, Kocór M, Krupnik S, Krygowska-Nowak N, Kwinta-Odrzywołek J, Worek B (2023) Trendy kształtujące polskie branże i kompetencje przyszłości. Polska Agencja Rozwoju Przedsiębior-czości, Warszawa

Świętek A (2022) Educational services in Poland during the COVID-19 pandemic. Stud Ind Geogr Comm Pol Geogr Soc 36(2):91–106. https://doi.org/10.24917/20801653.362.6

The Future of Jobs Report (2023) https://www.weforum.org/publications/the-future-of-jobs-report-2023/digest/. Last access 1 Apr 2024

The impact of globalisation on employment and the EU, Official Europarliament website (2023). https://www.europarl.europa.eu/topics/pl/article/20190712STO56968/wplyw-globalizacji-na-zatrudnienie-i-na-ue. Last access 1 April 2024

Toffler A (1998) Szok przyszłości. Wyd. Kurpisz, Poznań

Wang Y, Liu S (2016) Education, human capital and economic growth: empirical research on 55 countries and regions (1960–2009). Theor Econ Lett 6:347–355. https://doi.org/10.4236/tel.2016.62039

Wawrzonek A (2020) Education, labour market—spaces of common needs and goals. Wychowanie w Rodzinie t. XXIII(2/2020):259–274. https://doi.org/10.34616/wwr.2020.2.259.274

Wilsz J (2014) Implikacje globalizacji dla rynku pracy. Studia Ekonomiczne, Zeszyty Naukowe Wydziałowe Uniwersytetu Ekonomicznego w Katowicach. Wydawnictwo Uniwersytetu Ekonomicznego w Katowicach 197:57–67

WHO handbook for guideline on mental health at work (2022). https://iris.who.int/bitstream/handle/10665/363177/9789240053052-eng.pdf?sequence=1. Last access 1 April 2024

Zieliński J (2007) Uwarunkowania i założenia przemian edukacyjnych w XXI wieku. Edukacja wobec rynku pracy i integracji europejskiej. WSB, Dąbrowa Górnicza

Zioło Z (2020) The role of education in socio-economic and cultural development. Entrepreneurship Educ (Przedsiębiorczość-Edukacja) 16(2):7–24. https://doi.org/10.24917/20833296.162.1

Looking Back to See What's Forward?—Heritage Education in Geography—New Understanding and Future of Students Training

Agnieszka Świętek and Piotr Cybul

Abstract Despite changes in the understanding of heritage in recent years, researchers' attention is still focused on searching for digital tools to improve and make the heritage education process more attractive, responding to the needs of generations Z and Alpha, who live simultaneously with the world of technology. Authors' experience indicates numerous problems among students in interpreting the symbolism and meaning of heritage during theoretical, practical and field classes. The authors seek answers to two questions: How do young generations understand cultural heritage? What will young people incline to strive to preserve it in the future and why? The authors will present the results of research on the perception of heritage and education for it by: primary school pupils from the generation Alpha and tourism and geography students from generation Z, who may create the educational and economic dimension of cultural heritage in the future. In 2023/2024, qualitative research was carried out using the diagnostic survey method. The results were analysed using the semantic field method and a relationship grid. The research showed the perception of heritage and its education by young generations. Conclusions from the obtained research results may serve the future development of "heritage education in geographical education.

Keywords Heritage · Geography of heritage · Heritage education · Generation Alpha · Generation Z

A. Świętek (✉) · P. Cybul
University of the National Education Commission, Krakow, Poland
e-mail: agnieszka.swietek@uken.krakow.pl

P. Cybul
e-mail: piotr.cybul@uken.krakow.pl

© The Author(s), under exclusive license to Springer Nature Switzerland AG 2025

D. Piróg and A. Świętek (eds.), *Future-Ready Geography*, Key Challenges in Geography, https://doi.org/10.1007/978-3-031-81268-2_3

Introduction

Future-ready geography should be ready to prepare students and pupils for adult life in the future world, the shape of which is impossible to determine. It is developing dynamically, and in addition to the traditional axioms such as death and taxes, the third given in it is only change. This does not exempt teachers of geography and researchers of geographic education from the obligation to strive to prepare the young generation in the best possible way to understand and, in the future, responsibly create reality. Unless we share the fate of the dinosaurs or encounter some other catastrophic phenomenon straight out of Roland Emmerich's films, the near future will continue to be shaped by intensifying civilization trends: modernisation (the development of technology, information technology and digitization, based on artificial intelligence) and globalisation (Gidley 2010; Hajkowicz et al. 2012). The young generation will face global problems: increasing climate change, environmental degradation, mass migration, refugees, xenophobia, racial, linguistic and religious discrimination, military conflicts and terrorist activities (González et al. 2018). "Geography, better than any other discipline, can explore the spatial relationships of social and physical phenomena at diverse scales and suggest solutions to significant and persistent problems. Therefore, geography education (…) has a tremendous potential to provide the next generation with the understanding, attitudes, and behaviours required to work for solutions to global problems" (González et al. 2018: 4).

One of the key challenges for Europe in the twenty-first century is the cultural diversity of its people. The disappearance of objective and mental barriers to the flow of people, goods and services as a result of globalisation has resulted in the migration of cultural patterns and customs (Wierzbicki 2000). This raises questions about the future shape of European culture, the nature of European heritage and the role that geographic education can play in it. The postmodern phase in Western civilization has resulted in the liberalisation of social life: the decline of traditional values in favour of postmodern self-development values and post-materialist priorities, including emancipation, equality and autonomy (Inglehart and Norris 2004; Inglehart and Welzel 2006). Secularisation has triggered changes in consciousness (decline in religiosity) and social-institutional ones (de-churchization). There has been an increase in acceptance of pluralism of values and diversity of lifestyles (Beck and Beck-Gernsheim 2002). New "sub-national" types of differentiation, groups and social movements have emerged, individualism, the role of the individual, acceptance of "otherness" and diversity have increased (Ziółkowski 2000). Democratic legal relations have granted numerous freedoms to ethnic, minority and religious groups, further spurring migration (Suchocka and Królikowska 2014). In a number of traditional communities, however, the cultural dominance of the West raises concern (Świętek and Brusik 2022). After the terrorist attacks of September 11, 2001, in the United States and later, in several European cities (for instance, Madrid, London, Paris, Berlin, Brussels), the validity of the idea of multiculturalism has been questioned (Valentine and Harris 2016). The issues of cohesion and integration, including encounters with diversity and the significance of creating common identities, have gained particular importance (Bloch et al. 2013; Dwyer and Bressey 2008; Meer and Modood 2009). In the era of globalisation of cultural patterns, the fear of getting lost in the world, losing personal, national and regional identity, displaced by a global identity, has caused a renewed interest in cultural heritage and local issues. (Suchocka and Królikowska 2014; Świętek and Brusik 2022). A counterbalance to the processes of globalisation is the renaissance of regionalism, referred to as the "ethnic awakening" or "return to the roots" (Skorowski 2006). It reflects a longing for deep identity and a desire to belong to authentic regional communities, with their history, culture and nature (Suchocka and Królikowska 2014).

Under these circumstances, the components of geographic education should undergo changes: multicultural education is even more crucial, while regional education and heritage education, along with the redefinition of heritage itself, are gaining a new meaning and role. The current understanding of heritage is inclusive, encompassing all natural and cultural values, emphasising diversity, equality and human rights (Apostolopoulou et al. 2014; Karadeniz 2020; Pérez-Guilarte et al. 2023). Heritage education should serve not only heritage, but also the subject of education, which is a student. Heritage education is no longer understood in accordance with educational conservatism as propagating the belief that material heritage needs to be protected. Its new role is to support students in developing their own identity, functioning in a multicultural society and adhering to the principles of sustainable development.

Nowadays, heritage education is becoming the quintessence of education, understood as an organised activity aimed at providing the young generation with comprehensive development and preparing it for adult life, in which it will effectively draw on the achievements of previous generations, build on it and pass on what is valuable to subsequent generations. Modern concepts of education and heritage are closely related. It is impossible to separate them because they interpenetrate and have the same goal. It is for students to acquire knowledge, skills and develop pro-social and pro-environmental attitudes (Mendoza et al. 2015). "Our heritage is something which forms a part of our daily lives" (Pérez et al. 2010: 1319), therefore, according to the principle of utilitarianism, it is a subject of knowledge in various school subjects. In formal education in accordance with the Faro Convention, Art. 13: "heritage dimension is suggested to be included at all levels of education, not particularly as a core subject but through interdisciplinary actions" (Apostolopoulou et al. 2014: 68). The content of education and upbringing, in particular geography, draws on the achievements of various scientific disciplines dealing with the study of social and natural objects and phenomena (Cuenca-López et al. 2021). This heritage, together with the scientific achievements related to it, constitutes the core of the modern geography curriculum in schools. Heritage education also goes beyond formal education at school and also takes place through informal education (Pérez et al. 2010).

For the development of geographic education that prepares young people for the future world, as well as the future of heritage itself, the key questions are: What do the representatives of the young generation consider heritage, what values do they assign to it, and what are their attitudes towards heritage? as well as: What are the needs of the younger generation that should be taken into account when designing future geographic heritage education? The search for answers to these questions became the subject of the authors' literature and empirical research.

Chapter four presents the results of the research conducted among representatives of Gen Alpha and Gen Z. In the study, the authors applied the diagnostic survey using a questionnaire and interview research, as well as the visual method. The statistical method was used to analyse the data, while written answers were analysed using the semantic field method.

Heritage Education—The State of Research

Changes in the Understanding and Perception of Heritage

The traditional understanding of heritage is reflected in the 1972 UNESCO Convention, which defines it as: monuments and sites of exceptional universal historical, aesthetic and scientific value, deserving to be preserved and passed on from generation to generation (Harrison 2010). In the twentieth century, its economic value was recognised, as well as the dangers of its commercialization (Pérez-Guilarte et al. 2023). Actions taken towards heritage involved its protection, conservation and popularisation, using so-called heritage practices: language, culture, music, literature or clothing, which are mementos and carriers of memories of given communities (Harrison 2010). It was assumed that heritage would be "transferred" through traditional lifestyles and by means of formal education (Srivastava 2015). In the spirit of educational conservatism, the task of heritage education was to shape people's ideas about their past, to convey knowledge about heritage objects and the need to protect them. The goal was to pass on traditional values and patterns of behaviour to the young and potential members of society, so that it would achieve basic social conformity (Srivastava 2015). The presumption was that new elements of culture should be implemented slowly in order to be more easily assimilated, and "the transmission of traditions has sometimes to contradict the desire to initiate change." (Srivastava 2015: 336).

The contemporary understanding of heritage is marked by trends: on the one hand, the commercialization and globalisation of culture, and on the other, the desire to return to searching for one's own identity in the globalised world. Heritage is considered an essential source of international prestige as well as a profitable tourist activity (Graham et al. 2000). Visiting heritage sites has become a common habit of members of wealthier societies and a field of strong public interest (Harrison 2012). Not only has the relevance of heritage expanded, but it has also been reinterpreted in the wake of dynamic social changes: rapidly growing cultural diversity, ideas of equality, and human rights (Karadeniz 2020). Contemporary scientific research on heritage is based on two paradigms: classical and/or constructivist (Pérez-Guilarte et al. 2023). In the classical perspective, heritage assets are top-down designated by elites and experts and are a value per se that must be protected at all costs (Poulios 2014; Smith 2006; Pérez-Guilarte et al. 2023). In the constructivist perspective, also known as the "values-based approach", it is the individual social groups, based on their cultural patterns and worldviews, that decide which values and resources constitute their heritage (Pérez-Guilarte et al. 2023). It becomes a social construct and the succession process is marked by generational conflicts. According to the assumptions of constructivism, the centre of gravity in heritage research shifts from the study of heritage elements to the values assigned to them by people (Pérez-Guilarte et al. 2023). The presumptions of both paradigms intermingle in the social and political spheres. One common way of classifying heritage is to distinguish between "cultural" ("human-made") heritage and "natural" (non-human-made) heritage, although modern researchers note that this division is rather automatic and not accurate in all cases of objects (Harrison 2010). The heritage categories considered in international and national regulations are based on those applied by the United Nations Educational, Scientific and Cultural Organization (UNESCO), or the International Council of Monuments and Sites (ICOMOS), among others. They include the following as heritage:

1. "1. Tangible cultural heritage (buildings, historical sites, monuments and all kinds of archaeological, architectural, technological and scientific works made by human hands).
2. Movable Cultural Heritage (sculptures, paintings, manuscripts, coins, archaeological artefacts, etc.)
3. Immovable cultural heritage (archaeological monuments, sites, historic city tissues, etc.)
4. Underwater Cultural Heritage (shipwrecks, underwater ruins and cities)
5. Intangible Cultural Heritage (rituals, oral traditions, performing arts, etc.)
6. Natural Heritage (natural sites with cultural dimension, cultural landscapes, biological, physical and geological formations etc.)" (Karadeniz 2020: 42).

Heritage today encompasses all tangible and intangible effects of human activity and elements of the natural environment that those who are now living want to adopt (inherit) and preserve in memory. (Graham et al. 2000; Ashworth et al. 2007; Howard 2003). Tangible heritage is movable objects representing the material culture of a society and immovable objects, cultural landscapes, historic cities, or buildings (Ciobanu 2016). Intangible heritage includes elements of culture and folklore, from language, customs, traditional dance, music and costumes, to cuisine and popular culture events, characteristic and unique to specific geographic locations (Karadeniz 2020). According to the constructivist paradigm, the catalogue of tangible and intangible heritage can be considered open. It can include a variety of natural elements as well as intellectual and spiritual achievements of mankind. Ciobanu (2016) adds, for example, such elements of "intangible" cultural heritage as: oral traditions and expressions, performing arts, social practices, rituals and festive events, knowledge and practices concerning nature and the universe. The openness of the heritage set stems from its social nature: "it is constantly reproduced by communities and groups in response to their environment, their interaction with nature and their history, ensuring their sense of identity and continuity, thus promoting respect for cultural diversity and human creativity" (Ciobanu 2016: 104). Therefore, what can be considered as a heritage resource are those instruments, objects, artefacts and cultural spaces that particular communities, groups and even individuals consider to be significant carriers of a community's identity and memory (Towse 2022; Throsby 2001). The assessment of their validity is strongly evaluative, subjective and dependent on the perspective taken. What for some people is heritage may not be such for others, depending on their individual evaluative judgements based on personal, regional and national identity (Harrison 2010). Moreover, the recognition of a given resource as heritage is a dynamic process. A resource considered worthless may over time be recognised as valuable (Throsby 2001). According to the authors, generational differences are crucial to this assessment and its changes.

However, the complex: classical and constructivist, "discretionary" and "social" nature of heritage has its direct translation into hard legislative and market practice. Resources that are widely recognised as valuable are protected by law, so they have a greater chance of not getting destroyed and being "inherited" by the next generation. They have their own market value—goods and services are created from them, particularly in the tourism services market. The importance of heritage for the economy, the commercialisation of culture, as well as mass tourism and its consequences, is an area of considerable scientific research. (Towse 2022; Bowitz and Ibenholt 2009; Throsby 2001; Murzyn-Kupisz 2010).

Heritage Education

Like the understanding of heritage, the understanding of education underwent significant changes in the twentieth century. Thanks to personalism, a pupil became the subject of education and the aim was to develop their knowledge, understanding, skills, competences, attitudes, values and, as a result, their personality. Cognitive constructivism has become a new educational paradigm in the last forty years (Olusegun 2015). Constructivism provided "a theoretical basis for building detailed praxeological approaches to learning" (Dylak 2020: 64), based on "observations and scientific research on how people learn" (Michalak 2020: 100). It is an alternative to the criticised paradigm of objectivism (Atroszko 2018). Its foundation is the assumption that people construct knowledge and make meanings based on their own experiences (Bereiter 1994). A student does not receive knowledge from outside, but creates "interpretations of the world based on their own past experiences and interactions between themselves and the world" (Wendland 2011: 39). Such an assertion has changed thinking about learning and teaching and has had a significant impact on school curricula, learning outcomes, recommended teaching resources and educational methods (Klus-Stańska 2020). Modern constructivists accept, following Piaget, that the basis of the learning process is experience and reflection on that experience, which builds knowledge and understanding of the world (Olusegun 2015). They are constructed by the learners themselves, who actively create them (Piaget

1966). In the course of the learning process, information gained from new experiences is confronted with the learner's existing knowledge, experiences and perceptions, as well as with the learner's environment (Phillips 1995). The result is the reconstruction of a picture of a piece of reality in the learner's mind and the creation of their new knowledge. There is a relative consensus in the research community on the acceptance of the assumptions of constructivism and the validity of implementing its principles in school practice (Olusegun 2015; Gruender and Tobin 1991; Dylak 2020; Klus-Stańska 2020).

The changes in heritage and in education, described briefly, should translate into an evolution in the understanding of heritage education and the way it is implemented in school practice. Both the modern understanding of heritage, personalism and the constructivist paradigm advocate looking at heritage education through the lens of "values-based approach" and apply the "critical theory of heritage " (Pérez-Guilarte et al. 2023). According to them, young people should not be passive recipients and "protectors" of elements of heritage top-down considered significant, but constructors of their own components and understanding of it, as building blocks of individual identity. "*The focus shifts from the object to the subject that creates, understands, enjoys, and, in this process, recreates the heritage. Therefore, the emphasis is not so much on the element itself but on the value attributed to it*" (Pérez-Guilarte et al. 2023: 6173). The development of a young person's cultural identity, in the course of creating their own vision of cultural heritage, has both personal and social significance. A developed individual identity does not contradict the idea of a multicultural society; on the contrary, it is key to its proper functioning (Ustrzycki 2007; Suchocka and Królikowska 2014). Literature research has shown that both science and school practice are only at the beginning of this journey.

In 2005, the member states of the Council of Europe adopted the Convention on the Value of Cultural Heritage for Society (Council Convention 2005), stipulating that heritage education content in national school systems shall be present at all levels of education, not necessarily as a separate subject, but integrated into the teaching of existing subjects (Council Convention, Article 13 2005). In the majority of European countries, raising awareness and promoting the value and role of heritage is integrated into history and geography curricula (Apostolopoulou et al. 2014). In spite of this, research conducted several years ago is still concerned proving the validity of implementing elements of heritage education into formal schooling (Wu 2011; Pooten 2014; Méndez 2006; Alkateb 2013). The possibilities of integrating heritage education content into the teaching of different school subjects were analysed (Smith 2011; Alajmi 2013; Alpin 2007; Averill 2012; Apostolopoulou et al. 2014). The aspect of politicising education through the selection and implementation of heritage education content into schools was also considered (Okubo 2010; Abayao 2006). Issues of proper preparation of teaching staff to deliver this content were raised (Perez and Templanza 2012; Averill 2012). Considerable attention was paid to strategies and methods of heritage education (e.g. Pazzagua and Williams 2012; Smith 2011; Simşeket et al. 2012), however, only few works present the results of studies on the effectiveness of their implementation in school practice (Ocampo and Delgado 2014).

In the new geography education curricula, there is an emphasis on school projects concentrated on local issues, where students are expected to conduct research using GIS and ICT (Klonari and Tzoura 2011). The researchers therefore focused their attention on the application of cutting-edge technology in school heritage education (Ott and Pozzi 2011; Cat et al. 2013; Bustillo et al. 2015; Wang et al. 2023). Investigations included the virtual visualisation of historical objects in the form of the 3D model for learning about heritage objects (Bustillo et al. 2015), the use of digital maps designed by students using a variety of applications (e.g. ArcGIS Online, myHisto, Google Maps, Maps Tours), for their learning about historical and archaeological monuments in an urban environment (Apostolopoulou et al. 2014), or the use of Augmented Reality (AR) based on Serious Game (SG) and GPS-guided applications among children to learn about cultural heritage sites (Capecchi et al. 2024). Achille and Fiorillo (2022) prepared a compilation of examples of good practices of the use of numerous technological tools in heritage education at all educational stages.

It is only in the last few years that the fundamental issues of perception and values in school heritage have been revisited. Shimray (2019) compiled "Ways to Create Awareness on Cultural Heritage distinguishing school education among them. In a 2022 article, Achile and Fiorillo (2022) rightly noted that: "it is not enough to know how to use technology and information tools well; it is necessary to spread awareness to build a participatory and informed mindset. A change of this magnitude can only take place in educational institutions. Along the educational pathway, the concepts of 'cultural heritage, conservation, preservation, care' must be introduced, ending with the concept of 'common good'" (Achile and Fiorillo 2022: 2567). Kafadar's (2021) research, whose assumptions can be considered constructivist due to the possibility of students indicating for themselves how they understand heritage and what its elements are, has led to compelling conclusions. It has shown that younger students from the Gen Alpha see more elements of intangible heritage than their older peers, which the author explains by focusing on tangible heritage in school education in older years at school.

It showed that younger pupils of the Alpha generation perceive more elements of intangible heritage than their older peers, which the author explains by focusing on tangible heritage during school education in the upper years of schooling (Kafadar 2021). Considering awareness as a key condition for young people's willingness to protect cultural heritage, Karadeniz (2020) conducted a study of geography students' perceptions and awareness of universal values and significant cultural heritage sites in Turkey. This research showed that their awareness of heritage values is moderate, and the author postulates the need for further research and the necessity to increase the quality of student education in this matter (Karadeniz 2020). The studies on the "values-based approach" in heritage education among young generations postulated by Pérez-Guilarte et al. (2023), have not yet been conducted.

A New Generation as a Challenge for Heritage Education

Designing future geographic heritage education requires looking at the youngest generation of students. According to the Howe and Strauss model (1991), generational change occurs approximately every 20 years, and some researchers believe that technological advances have shortened it to only 15 years (Thompson 2018). The current young people belong to two generations: Z and Alpha. Generation Z includes students and high school pupils born between 1990 and 2010, while the youngest Alpha generation are children born after 2010. (Thompson 2018). Their most distinctive feature is the fact that they are the first generations for whom technology and social media are a parallel area of their life (Ziatdinov and Cilliers 2021; Thompson 2018; Nagy and Kölcsey 2017). Numerous scientific papers have been written on Generation Z, while the Alpha cohort is not yet sufficiently examined, due to its relatively young age (Ziatdinov and Cilliers 2021; Amrit 2020; Thompson 2018). Existing studies indicate that the functioning of Alpha is even more strongly influenced by access to technology and information (Ziatdinov and Cilliers 2021; Tootell et al. 2014). Compared to Gen Z, the effects of modernisation are in their case amplified (Thompson 2018; Nagy and Kölcsey 2017; McCrindle and Fell 2020). Research by Apaydin and Kaya (2020) shows that Alphas have high levels of perception and numerical intelligence. The time they spend playing in front of their smartphone, iPad or laptop screens makes them more tech literate and the virtual environment is a natural part of their reality (McCrindle and Fell 2020). Thanks to their strong interactivity (van Dijk 2010), representatives of both young generations are able to learn independently and quickly acquire the necessary information (Kubów 2019; van Dijk 2010). In the learning process, Gen Alpha, more strongly than Gen Z, expect the application of visual, auditory and kinesthetic methods, which correspond to their habits developed through the use of technology and the Internet from an early age (Apaydin and Kaya 2020). Online content is dominated by visual elements (videos, images, memes, emoticons), easily digestible and ambiguous. Texts created by so-called average internet users are short and written in internet slang (Zabawa 2011). They are characterised by a high degree of interactivity, allowing the audience to co-create them by adding comments, both on the main text and on other users' entries (Zabawa 2011).

The consequence of young people's habit of consuming large amounts of fast and often unprofessional information is their multitasking, short attention span and superficial understanding (McCrindle and Fell 2020). It may result in surface approach to learning, which Dejene et al. (2018) defines as memorising or reproducing facts, rather than engaging in active research, linking facts to conclusions, reflecting on the author's intentions and integrating content with personal experience. In line with the findings of Abedin et al. (2013), students with such an attitude do not bother to understand the meaning of the content they read (Abedin et al. 2013). Apostolopoulou et al. (2014) argue that the challenge for future academics will not be to teach how to acquire information, but to translate, interpret and add value to it, as well as to improve literacy skills.

The globalisation and symbiosis of the Z and Alpha generations with technology is not only reflected in their cognitive development and the way they learn, but also in their social development and attitudes towards heritage. Existing subjects of social control, such as the family, the school or the Church—based on direct interaction and personal authority, using classic mechanisms of socialisation—are losing importance, in favour of the control of the mass media, popular culture and the market for consumer goods (Ziółkowski 2000). These support a unified global identity of their young recipients and promote a consumer lifestyle among them (Suchocka and Królikowska 2014). Apaydin and Kaya's (2020) study of pre-school teachers identifies, in addition to technology addiction, a tendency towards violence and egocentricity as well as poor social intelligence as the main problems faced by children of a few years of age, as a consequence of limited interpersonal contacts in the real world (Apaydin and Kaya 2020).

On the other hand, the Internet offers the possibility of unlimited contacts that remove age, class or national divisions (Pyżalski 2019). Young people are open to multiplied interactions and social contacts, willing to collaborate, share knowledge and solve problems online (Baron-Polańczyk and Nowak-Łojewska 2022). The easy access to information on global issues and their popularisation on social media results in their rapid dissemination among the Z and Alpha generations. Examples of fashionable positive attitudes created in this way include healthy, sporty lifestyles, vegetarianism, sensitivity to the fate of animals, or environmental attitudes that are crucial for the natural heritage. Research by Shutaleva et al. conducted among young people (2022) showed that social media, rather than school or family, are the strongest generators of environmental knowledge and behaviour, while bloggers, musicians, athletes and activists are authority figures (Shutaleva et al. 2022). That leads to digital initiatives such as Youth Strike for Climate, which is an international movement of thousands of young people protesting against the inactivity of politicians towards global warming and climate change (Poot and Bauwens 2023).

The development of young people in an era of "global ecumenism", homogenisation and commercialisation of culture, has implications for their understanding of cultural heritage (Suchocka and Królikowska 2014). In social media, they learn about values different from traditional ones and diverse lifestyles (Beck and Beck-Gernsheim 2002), which increases their level of acceptance of "otherness" and diversity, as well as their willingness to emphasise their own individualism (Ziółkowski 2000). They have the possibility to appear in much larger, international and culturally diverse groups and to create their own image freely. Creating one's original E-identity, impressive to other Internet users, is a crucial source of motivation for Gen Z and Alpha. It prompts them to give accounts of their lives online, and sometimes to share with others: "things so weird as even funny" (Prościak 2019: 229), often for profit. Young people are strongly threatened by the unhealthy pursuit of fame and the cult of money. They are extremely pragmatic in their everyday decisions. They make them with a view to their own benefit and/or money rather than to values (Czerniawska 2016). Young people can experiment with their own identity, take on the roles of others, identify with game characters and experience their emotions during fictional events (Baron-Polańczyk and Nowak-Łojewska 2022). They are subject to the influence of multiple, often conflicting sources of information, which makes it difficult for them to achieve a stable image of their "self" (Nikitorowicz 2009). The issue of adolescent identity is widely discussed in the literature (Suchocka and Królikowska 2014). "The construction of identity uses as building blocks material from history, geography, biology, from institutions of production and reproduction, from collective memory and personal dreams, the apparatuses of power and religious revelation. However, these materials are processed by individuals, social groups and societies" (Castells 2008: 23). Due to liberalisation, young Europeans choose between identities disseminated by the media. A homogenised version of global culture offers them uniform cultural patterns, a unified lifestyle and a superficial identity (Suchocka and Królikowska 2014). It does not require the effort of searching for one's roots and insight into oneself, it is easy and "fashionable" and therefore often accepted uncritically (Suchocka and Królikowska 2014). "There is unification of the identity of the younger generation of teenagers, with the emerging category of the 'global teenager', whose identity is far less shaped by values (…). The phenomenon of the so-called 'identity to rent' is emerging, without roots, foundations, a sense of belonging, everywhere and nowhere 'at home'" (Suchocka and Królikowska 2014: 75). The selection of unstable identities does not meet the psychological and social needs of young people. There is a growing phenomenon of tribalism, or "temporary empathic communities (new tribes), within which people with no permanent identity seek (…) emotional identification with others in order to establish some kind of relationship (Burszta and Dudzikowa 2007: 221).

Research

To answer the research questions and understand how representatives of the younger generation perceive heritage, what values they assign to it and what their attitudes are in this respect, the authors conducted qualitative studies. They employed a diagnostic survey method using questionnaires and interview research, as well as a visual method to obtain data. A statistical method was used to analyse the data and the written responses were analysed using the semantic field method, using a relationship grid. The method is based on the assumptions proposed by Robin (1980) and developed by Kłosiński (1994). The study distinguished the following networks: equivalents, terms, associations, oppositions, actions of the subject and actions towards the subject. A separate analysis of the semantic field "heritage was performed for each group of respondents. Due to the respondents' statements being too short and often referring to several elements, the authors modified the method by grouping or separating the statements of the participants in the study on the basis of matching the content to the individual equivalents. On the other hand, respondents' written statements on education were often in the form of single entries, so the decision was made to conduct a word-frequency analysis. Statistical analysis of the text was performed using the online text tool, which was preceded by the removal of stop-words at the pre-processing stage.

The research group comprised 48 representatives of the Alpha generation, 4th grade pupils of primary school and 67 representatives of Gen Z, including 52 tourism and recreation students and 15 geography students. The age of the respondents and their potential impact on heritage and heritage education in the future were considered as criteria for selecting participants for the study. Pupils from the 4th year of primary school were selected for the study as some of the oldest representatives of the Alpha generation in Poland. The choice of this age group is justified by the provisions of the core curriculum for Grade 4, since, according to its requirements, students at this stage should be able to:

- evaluate the landscape in terms of its beauty and the cultural and natural heritage of their "little homeland"
- indicate the locations of protected areas, natural monuments and historic buildings in the immediate vicinity, and justify the need to protect them (Journal of Laws 2017, item 356)

Tourism and recreation as well as geography students, on the other hand, were selected because they belong to the Gen Z. The curricula of courses with heritage-related content were also included. It was also crucial that tourism and recreation students, once they obtain their diploma, will be qualified to work in travel agencies, offices and state institutions dealing with tourism. In this way, these people can influence the creation of directions for tourism development and, in line with constructivist theory, will be able to define the elements considered as heritage. Those who will land jobs in travel agencies, on the other hand, will be responsible for creating and selling tourism offers, including those aimed at schoolchildren as part of school tourism.

A group of geography students, teaching specialisation, also representatives of generation Z, who will be qualified as geography and science teachers after graduation, were also selected for the study.

The authors developed an interview questionnaire consisting of 5 questions and demographics. The questions in the questionnaire were formulated in such a way that the youngest participants in the study (9–10 years old) would have no problem understanding them, but also that the students (around 22–24 years old) would not find them too trivial. The word "heritage was not used explicitly in the questions, not to suggest to the respondents the meaning and elements of heritage they were supposed to indicate. An additional assumption of the authors in designing the questions was to minimise their number taking into account the limited perception time of the younger generations (McCrindle and Fell 2020). To limit superficial responses, the time for working on the questionnaire was not limited. This allowed each respondent to give answers at their own pace and to focus on the instructions. This was particularly relevant for the task using the graphic method, which aimed to identify elements considered as heritage by representatives of the Alpha and Z generations. In this task, an image depicting the landscape characteristic of the region in which the survey participants live (Lesser Poland, Poland) was generated using artificial

Fig. 1 An image showing heritage elements specific to the region in which the respondents live. Drawn up by the authors using AI

intelligence. Then, with collage techniques, the authors supplemented the image with elements representing the other heritage categories mentioned by Karadeniz (2020: 42) (Fig. 1). The image included both cultural and natural heritage elements, taking into account tangible and intangible heritage sites. Each heritage category was represented in the picture by at least one object. Participants indicated in the image all the elements that they thought were worth preserving for future generations. Each participant was also given the possibility to draw or write in elements that, in their opinion, were missing from the picture and to cross out elements that, in their view, are not worth preserving for future generations. This possibility corresponds with constructivist assumptions.

In order to determine the values that the respondents were guided by when marking or adding particular heritage items, they were asked to explain these decisions in writing. Another aim of the survey was to verify the attitudes of the representatives of generation Alpha and Z towards particular elements of heritage. Attitudes were verified on the basis of willingness to financially support specific activities and initiatives (Fig. 3). Ten elements representing different categories of cultural and natural heritage were included in the table. The descriptions are accompanied by icons symbolising the given objective, in order to facilitate the perception of the youngest participants of the study, who are used to the graphic form of information with which they have daily contact on the Internet. Respondents were given a budget of PLN 100 (the equivalent of around €25), which they could use to support any or all of the initiatives listed.

Results

The analysis of the results enabled the identification of the resources considered as heritage components by representatives of the Alpha and Z generations (Fig. 2), and the values they ascribe to them. Both generations most frequently marked the Royal Castle in Krakow (95%) as a most significant landscape feature worth preserving for future generations. The analysis of the semantic field showed that pupils and students unanimously showed an association of the Wawel Castle with the symbol of the city of Krakow which should be "known" and "remembered". In this case, the differences between the generations were only shown by the network of terms. Representatives of Gen Z described the castle as "the most important monument", while the younger generation of Alpha referred to it as "nice", "old" and "big".

The analysis of the other elements marked in the image showed much greater intergenerational differences in the perception of heritage between Alphas and Zetas. Representatives of Gen Z mostly noticed elements of intangible culture. This is evidenced by the most frequent indication of folklore groups (84%) as a key element in the image of the region, which they linked to "customs", "folklore", "folk music", "folk dance" or "folk costume" and "dialect". Gen Alpha less frequently indicated this element (42%) as crucial and those who identified it did not explain the motive for their choice.

Highly significant differences were noted in the case of natural heritage elements. Gen Alpha indicated them as important and even drew their own suggestions for objects in this category (sun, air, birds and other animals). Geography students (with qualifications to teach science), on the other hand, regarded natural elements as

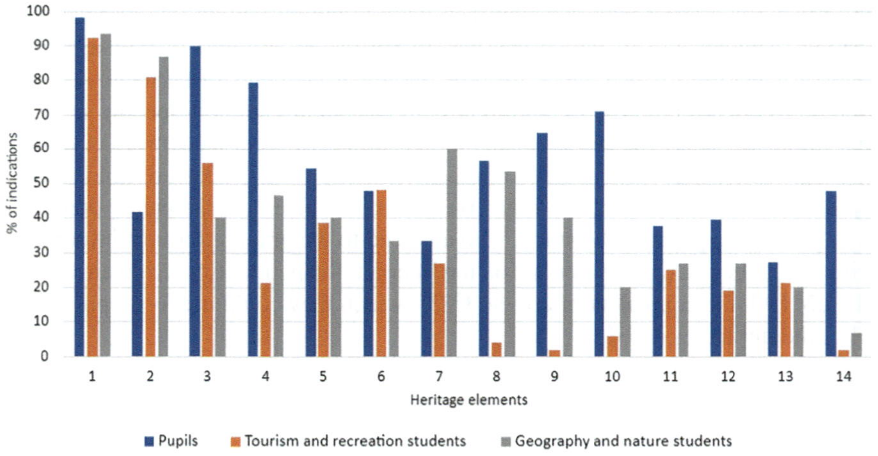

Fig. 2 Indications of individual heritage elements by respondents 1—Royal castle; 2—Folklore group; 3—monument to the Wawel Dragon; 4—sheeps; 5—monument to a famous poet; 6—Lajkonik—unofficial symbols of the city; 7—rural buildings; 8—mountains; 9—forest; 10—river; 11—archaeological excavations; 12—historic building; 13—painting; 14—industrial city

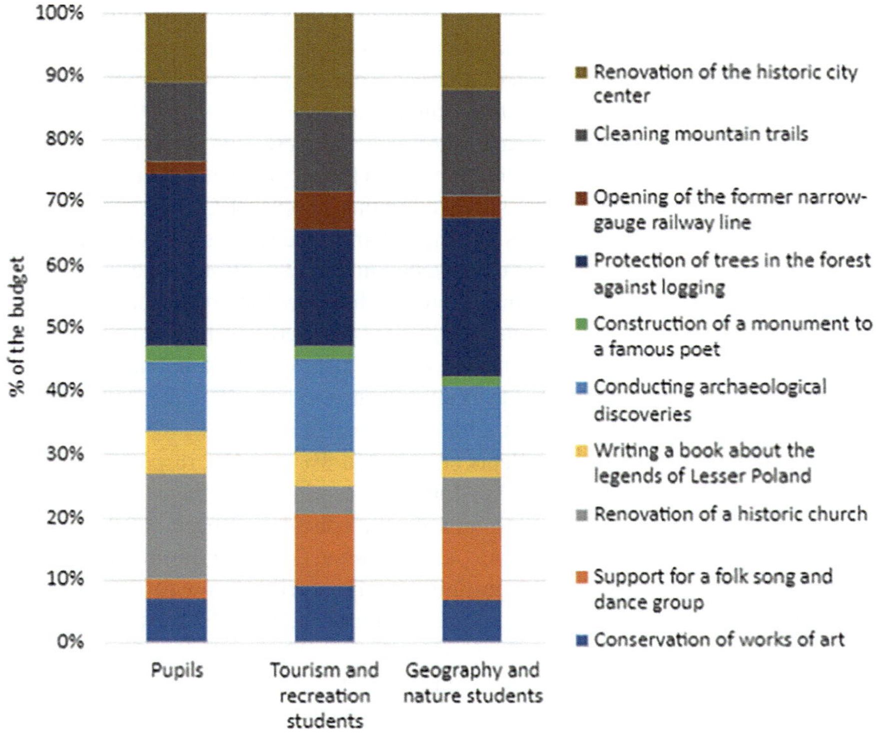

Fig. 3 Financial support for activities and initiatives related to various elements of heritage

heritage to a lesser extent, and tourism and recreation students hardly indicated them as crucial at all. The analysis of the semantic field showed quite different values attributed to natural heritage by representatives of the Alpha generation compared to Gen Z. Representatives of the Alpha generation mainly paid attention to the practical actions of the subject, proving their pragmatism as described in the literature (Czerniawska 2016). They regarded the forest as a "source of oxygen", a "possibility to breathe", or "fuel". Similarly, they viewed water through the prism of "beverage", which is "essential to life". They saw animals as a "source of food" ("meat"), "material for clothing", and as "friends of humans". Gen Z, on the other hand, linked nature with "ecology" and the "clean environment". The practical approach of the Alpha generation is also confirmed by pointing to the industrial city, which the students regard as "workplaces", while the students hardly considered this element as relevant to heritage at all. In times of widespread secularisation of society, the authors' attention was drawn to the religious heritage object—such as the church or the cross—which was often drawn by the Alpha generation. Historical churches can indeed be considered a secular, landscape element of Krakow's heritage, especially St Mary's Church. However, students who chose to include this element in their explanations linked it to "prayer" and "faith".

The analysis of respondents' attitudes indicated that both generations would be most willing to support natural heritage initiatives (Fig. 3). The majority of the budget (24%) was spent by them on protecting trees from logging and cleaning mountain trails (14%). This attitude contradicts the significant heritage elements they indicated. This is particularly evident in the case of tourism and recreation students, who rarely indicated essential elements of natural heritage in the picture, yet allocated the largest budget to them. According to the surveyed young people of Gen Z, recognising a site as a crucial heritage element is not equivalent to the willingness to fund it.

In contrast, Gen Alpha showed consistency in the perception of heritage and the attitudes displayed. For instance, they were much less willing to financially support a folk song and dance group, as their responses showed that they do not consider it as part of their heritage. Students allocated only 3% of their budgets for this purpose, with students in both fields of study contributing 12% each. In the case of the Renovation of a historic church initiative, students allocated 17% of the budget, while geography and tourism students contributed 8% and 4%, respectively. These results clearly correspond to the Gen Alpha's perception of heritage.

The research made it possible to identify the educational needs of the younger generation, which should be taken into account when designing future geographical heritage education. Recurring suggestions for appropriate ways to teach heritage were: Trips, Movies, Presentation, Teacher, Lessons, Fun, Exercises, Listening, Telling, Writing, Photos, Internet, Interesting (Fig. 4). The most common word in the responses from all groups was "excursions", and it appeared particularly frequently in the statements of geography students. In addition to excursions, pupils' responses mainly indicated to "movies", "teacher", "lessons", "fun" and "exercises". Students used these words to a lesser extent in their responses.

A significant discrepancy between the expected ways of teaching heritage education by pupils and students-potential teachers is shown, which can serve as a guideline when designing geographical education at school and university since they indicate the expectations and educational needs of both generations.

The results of the conducted research clearly indicate that school education plays an essential role in acquiring heritage knowledge for all generations (Fig. 5). It is also worth noting that for the Alpha generation, even the internet is not a crucial source of knowledge about heritage as school, family and excursions. The results obtained differ from those presented by Shutaleva et al. (2022), indicating a lesser role of schooling and family than social media on young people's attitudes.

Conclusions and Discussion on a New Look at Heritage Education in Geography

Literature research on the youngest generations Z and Alpha identified several significant features for designing future heritage education. The first of these, recognised by the researchers, is the need to use technology and social media in the learning

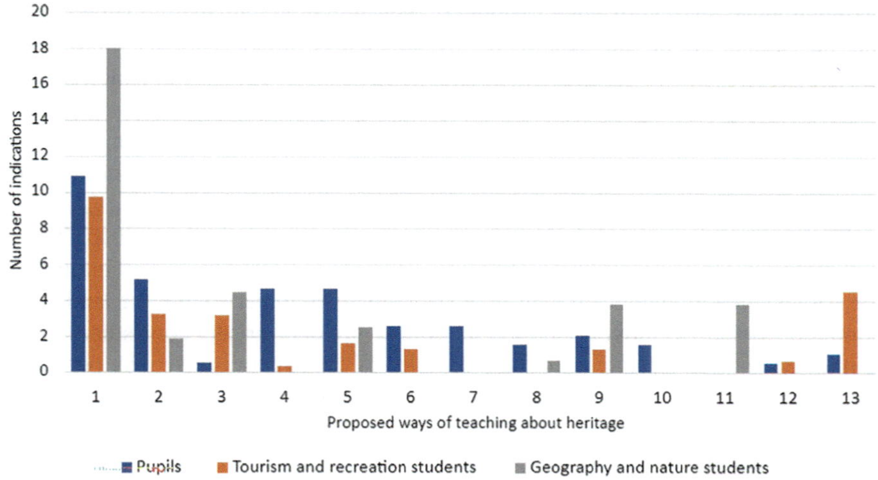

Fig. 4 The most commonly used terms in proposed ways of teaching about heritage in school education 1—trips; 2—presentation; 3—teacher; 4—lessons; 5—fun; 6—exercises; 7—listening; 8—telling; 9—writing; 10—photos; 11—internet; 12—interesting

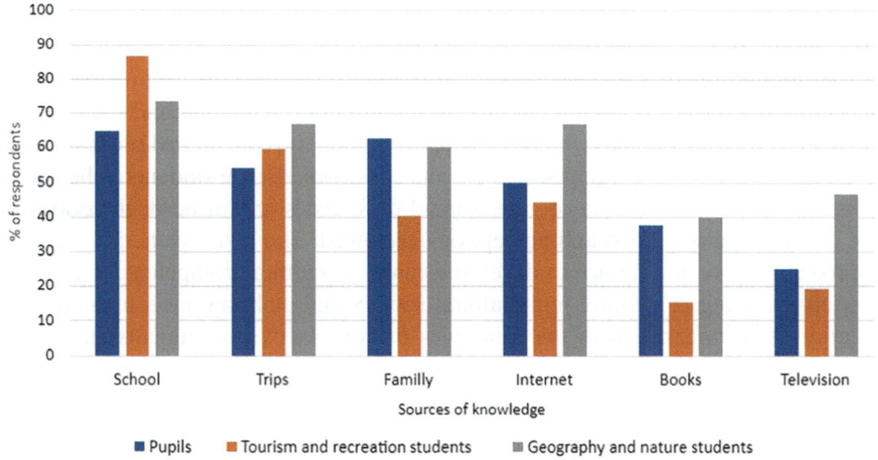

Fig. 5 Gen Z and Alpha's sources of knowledge on heritage elements

process. Such attempts and proposals for teaching solutions are already being implemented in education. At the same time, it was concluded that their use is crucial but that it is only a means to achieve the stated goals of heritage education, which is the common good. A significant conclusion from both constructivist theory of cognition and young people's experiences with interactive online content is their high potential independence in creating heritage elements and giving them value. Heritage understood as a social construct will therefore gain value online in the future. Another essential characteristic of young people is their potential tendency to superficially analyse elements of the surrounding reality, which may influence them to make unreflective assessments of cultural and natural heritage sites. This phenomenon may be exacerbated by the progressive individualism and pragmatism of young people identified in the literature. Their focus on the material aspects of life, consumerism and the disappearance of traditional values may significantly alter their perception of the importance of intangible heritage elements.

The results of the empirical study partly confirmed the findings of the literature research, albeit not fully. The heritage elements indicated by representatives of the Alpha generation confirm their greater pragmatism in perceiving the landscape, compared to representatives of Gen Z. They also prove the greater consistency of the attitudes displayed. In line with the principles of constructivism, young people of the Alpha generation were far more likely to create their own catalogue of heritage elements, adding or deleting them. Students, on the other hand, were less willing to create their picture of heritage. Despite having a much broader knowledge of the subject, they were laconic in both added and deleted elements. Moreover, they underestimated the value of the natural heritage of their region and appreciated more the sphere of intangible cultural heritage, related to the folklore and tradition of the region. There were also differences between representatives of the Alpha and Z generations in their didactic preferences. While all respondents appreciated the value of excursions in learning about heritage, other active forms of learning were also very important for younger students, yet teachers also played a major role.

Designing effective heritage education in the future must take into account the generational changes occurring and focus attention on the learner and their needs. What is worrying from the point of view of evaluating the educational process so far is the narrowing of the catalogue of elements considered by the students as heritage. This may be indicative both of generational differences and of heritage education not being properly guided and perceptions narrowed. The consequence of such an approach may be the failure to develop a range of cultural competencies that are needed for living in a future multicultural society and building one's own cultural identity. In accordance with the principles of constructivism, students should be allowed to choose for themselves the elements of heritage that carry their identity and to be more flexible in the educational strategies they adopt. The teacher's role in this process should be to explain the meaning and to support the deployment of the role of the different elements of heritage so that the pupil is able to attribute value to them independently.

Tailor-made heritage education will help young people develop their own cultural identity, providing them with a sense of belonging and an axiological system (attitudes and values) to which they will be able to refer in ambiguous situations (Nikitorowicz 2009).

References

Abayao LE (2006) Articulating indigenous people's culture in education. In: Saad I, Champagne D (eds) Indigenous education and empowerment: international perspectives. Altamira Press, Lanham, pp 179–187

Abedin F, Jaafar Z, Husain S, Abdullah R (2013) The validity of assist as a measurement of learning approach among MDAB students. Soc Behav Sci 90:549–557. https://doi.org/10.1016/j.sbspro.2013.07.125

Achille C, Fiorillo F (2022) Teaching and learning of cultural heritage: engaging education, professional training, and experimental activities. Heritage 5(3):2565–2593. https://doi.org/10.3390/heritage5030134

Alajmi FA (2013) Teaching cultural heritage through craft in Kuwaiti primary schools for girls

Alkateb M (2013) Non-traditional education using cultural heritage: a case study from Syria. Int J Educ Art 9(2):189–204. https://doi.org/10.1386/eta.9.2.189_1

Amrit KJ (2020) Understanding generation Alpha. https://osf.io/d2e8g/download. Accessed 08 Feb 2024

Apaydin Ç, Kaya F (2020) An analysis of the preschool teachers' views on alpha generation. Eur J Educ Stud 6(11):124–141. https://zenodo.org/records/3627158

Aplin G (2007) Heritage as exemplar: a pedagogical role for heritage studies in values education. Environmentalist 27(3):375–383. https://doi.org/10.1007/s10669-007-9053-4

Apostolopoulou A, Carvoeiras LM, Klonari A (2014) Cultural heritage and education. Integrating tour maps in a bilateral Project. Eur J Geogr 5(4):67–77

Ashworth GJ, Graham B, Tunbridge JE (2007) Pluralising pasts: heritage, identity and place in multicultural societies. Pluto Press, London

Atroszko B (2018) Konstruktywizm jako źródło inspiracji dla rozwoju współczesnej edukacji nauczycieli. In: Reviewed proceedings of the interdisciplinary scientific international conference for PhD students and assistants QUAERE 2018; June 27–29. Hradec Králové, Czechy, pp 1224–1231

Averill R (2012) Reflecting heritage cultures in mathematics learning: the views of teachers and students. J Urban Math Educ 5(2):157–181

Baron-Polańczyk E, Nowak-Łojewska A (2022) Pokolenie Alfa w szkole. Działania projektowo-wdrożeniowe z wykorzystaniem ICT. Problemy wczesnej edukacji 2(55):94–105. https://doi.org/10.26881/pwe.2022.55.07

Beck U, Beck-Gernsheim E (2002) Individualization. Sage, London

Bereiter C (1994) Constructivism, socioculturalism, and Popper's World 3. Educ Researcher 23(7):21–23

Bloch A, Neal S, Solomos J (2013) Race, multiculture and social policy. Palgrave Macmillan, Basingstoke

Bowitz E, Ibenholt K (2009) Economic impacts of cultural heritage—research and perspectives. J Cult Herit 10(1):1–8

Burszta W, Dudzikowa M (2007) Uczyć się od idola. O istotnym źródle wiedzy potocznej uczniów. In: Dudzikowa M, Czerepaniak-Walczak M (eds) Wychowanie. Pojęcia-Procesy-Konteksty. interdyscyplinarne ujęcie. GWP, Gdańsk, p 228

Bustillo A, Alaguero M, Miguel I, Saiz JM, Iglesias LS (2015) A flexible platform for the creation of 3D semi-immersive environments to teach Cultural Heritage. Digit Appl Archaeol Cult Herit 2:248–259

Capecchi I, Bernetti I, Borghini T, Caporali A, Saragosa C (2024) Augmented reality and serious game to engage the alpha generation in urban cultural heritage. J Cult Herit 66:523–535

Castells M (2008) Siła tożsamości. PWN, Warszawa, p 23

Cat K, Mooney J, Riley L, Howard-Wagner D (2013) Teaching through story mapping. Int J Technol Knowl Soc 9(3):73–84

Ciobanu C (2016) Space and time perception and the Geopark's communities. From mythical geography to heritage interpretation. Int Rev Soc Res 6(2):98–106

Council of Europe Framework Convention on the Value of Cultural Heritage for Society (2005) Council of Europe Treaty Series—No. 199
Cuenca-López JM, Martín-Cáceres MJ, Estepa-Giménez J (2021) Teacher training in heritage education: good practices for citizenship education. Humanit Soc Sci Commun 8(62):01–03. https://doi.org/10.1057/s41599-021-00745-6
Czerniawska M (2016) Postawy „pragmatyczne" versus postawy „wrażliwości społecznej" i ich aksjologiczne uwarunkowania. Prakseologia 158(1):137–160
Dejene W, Bishaw A, Dagnew A (2018) Preservice teachers' approaches to learning and their teaching approach preferences: secondary teacher education program in focus. Cogent Educ 5:1–15. https://doi.org/10.1080/2331186X.2018.1502396
Dwyer C, Bressey C (2008) New geographies of race and racism. Ashgate, Aldershot
Dylak S (2020) Behawioryzm i konstruktywizm wobec człowieczych czynów, wiedzy i moralności… ku (nie)podzielanej edukacji. Problemy Wczesnej Edukacji 4(51):60–74
Faro Convention. https://www.coe.int/en/web/conventions/full-list?module=treaty-detail&treaty num=199. Accessed 19 Jun 2024
González R, Bednarz S, Demirci A (2018) Why geography education matters for global understanding? In: Demirci A, González R, Bednarz S (eds) Geography education for global understanding. Springer, pp 3–13
Graham B, Ashworth GJ, Tunbridge JE (2000) A geography of heritage. Routledge, London
Gruender CD, Tobin K (1991) Promise and prospect. Sci Educ 75(1)
Harrison R (2010) What is heritage? In: Harrison R (ed) Understanding the politics of heritage. Manchester University Press, Manchester, pp 5–42
Howard P (2003) Heritage: management, interpretation, identity. Methuen, London
Howe N, Strauss W (1991) Generations: the history of America's future, 1584 to 2069. William Morrow & Company, New York
Kafadar T (2021) Cultural heritage in social studies curriculum and cultural heritage awareness of middle school students. Int J Prog Educ 17(2):260–274
Karadeniz CB (2020) Assessment for awareness and perception of the cultural heritage of geography students. Rev Int Geogr Educ (RIGEO) 10(1):40–64. https://doi.org/10.33403/rigeo.640722
Klonari A, Tzoura M (2011) The use of GIS for understanding geographical and environmental concepts-creating teaching material. In: Jeckel T, Koller A, Donert K, Vogler R (eds) Learning with GI 2011 implementing digital earth in education. Wichman Verlag, Germany, pp 38–47
Klus-Stańska D (2020) Konstruktywizm edukacyjny—niejednoznaczność, kontrowersje, dylematy, Problemy wczesnej edukacji 4(51):7–20. https://doi.org/10.26881/pwe.2020.51.01
Kłosiński M (1994) Semantyczna analiza pojęć „bezrobocie" i „bezrobotny" („bezrobotni") w wypowiedziach prasowych. Kul-tura i Społeczeństwo 38.3:151–162
Inglehart R, Norris P (2004) Sacred and secular. Religion and Politics Worldwide, Cambridge. https://doi.org/10.1017/CBO9780511894862
Inglehart R, Welzel Ch (2006) Modernization, cultural change and democracy. The Human Development Sequence, Cambridge. https://doi.org/10.1017/CBO9780511790881
Gidley J (2010) Globally scanning for "mega-trends of the mind": potential futures of futures thinking. Futures J Policy Plan Futures Stud 42.10:1040–1048. https://doi.org/10.1016/j.futures.2010.08.002
Hajkowicz SA, Cook H, Littleboy A (2012) Our future world: global megatrends that will change the way we live. The 2012 Revision. CSIRO, Australia
Kubów M (2019) Pokolenie Z—dzieci cyfrowej rewolucji. Digital and more. https://digitalandmore.pl/pokolenie-z-dzieci-cyfrowej-rewolucji/. Accessed 01 Mar 2024
McCrindle M, Fell A (2020) Understanding generation Alpha. McCrindle Research
Meer N, Modood T (2009) The multiculural state we're in: Muslims, 'multiculture' and the 'civic rebalancing' of British multiculturalism. Polit Stud 57:473–497
Méndez G (2006) Using students' cultural heritage to improve academic achievement in writing. Multicult Perspect 8(4):29–38

Mendoza R, Baldiris S, Fabregat R (2015) Framework to heritage education using emerging technologie. Procedia Comput Sci 75:239–249. https://doi.org/10.1016/J.PROCS.2015.12.244
Michalak R (2020) Konstruktywistyczna perspektywa wczesnej edukacji przyrodniczej. Problemy Wczesnej Edukacji 4(51):99–113
Murzyn-Kupisz M (2010) Podmioty na rynku dziedzictwa kulturowego. Studia Regionalne i Lokalne 3(41):61–80
Nagy Á, Kölcsey A (2017) Generation Alpha: marketing or science. Acta Educationis Generalis 7(1):107–115. https://doi.org/10.1515/atd-2017-0007
Nikitorowicz J (2009) Edukacja regionalna i międzykulturowa. Wydawnictwa Akademickie i Profesjonalne, Warszawa, p 371
Ocampo M, Delgado P (2014) Basic education and cultural heritage: prospects and challenges. Int J Humanit Soc Sci 4(9):201–209
Okubo YY (2010) Heritage: owned or assigned? The cultural politics of teaching heritage language in Osaka, Japan. Crit Asian Stud 42(1):111–138. https://doi.org/10.1080/14672710903537530
Olusegun S (2015) Constructivism learning theory: a paradigm for teaching and learning. J Res Method Educ 5(6):66–70. https://doi.org/10.9790/7388-05616670
Ott MM, Pozzi FF (2011) Towards a new era for cultural heritage education: discussing the role of ICT. Comput Hum Behav 27(4):1365–1371. https://doi.org/10.1016/j.chb.2010.07.031
Pazzagua G, Williams C (2012) Bring food and culture to the classroom. J Sch Health 82(12):577–580. https://doi.org/10.1111/j.1746-1561.2012.00739.x
Perez MV, Templanza MR (2012) Local studies centers: transforming history, culture and heritage in the Philippines. IFLA J 38(4):297. https://doi.org/10.1177/0340035212465033
Pérez RJ, Cuenca López JM, Ferreras-Listán M (2010) Heritage education: exploring the conceptions of teachers and administrators from the perspective of experimental and social science teaching. Teach Teach Educ 26:1319–1331. https://doi.org/10.1016/j.tate.2010.01.005
Pérez-Guilarte Y, Gusman I, Lois González RC (2023) Understanding the significance of cultural heritage in society from preschool: an educational practice with student teachers. Heritage 6:6172–6188. https://doi.org/10.3390/heritage6090324
Phillips DC (1995) The good, the bad, and the ugly: the many faces of constructivism. Educ Res 24(7):5–12
Piaget J (1966) Studia z psychologii dziecka. Wydawnictwo Naukowe PWN, Warszawa
Poot F, Bauwens J (2023) 'Like the oceans we rise': news frames on youth for climate. Young 31(2):107–123. https://doi.org/10.1177/11033088221115964
Pooten PM (2014) K+12, a seemingly inferior & anti-people program. Northern Dispatch Nordisweekly. http://www.nordis.net/?p=12952. Accessed 06 Feb 2024
Poulios I (2014) Discussing strategy in heritage conservation. J Cult Herit Manage Sustain Dev 4:16–34
Prościak M (2019) Między cyfryzacją, gamingiem a nauką—o motywacjach pokolenia XD. In: Niemierko B, Szmigel K (eds) Znaczenie diagnostyki edukacyjnej dla procesu kształcenia. Polskie Towarzystwo Diagnostyki Edukacyjnej, Kraków, pp 207–230
Pyżalski J (2019) Cyfrowa pedagogika medialna. In: Kwieciński Z, Śliwerski B (eds) Pedagogika. Podręcznik akademicki. Wydawnictwo Naukowe PWN, Warszawa
Robin R (1980) Badanie pól semantycznych: doświadczenia Ośrodka Leksykologii Politycznej w Saint-Cloud. In: Głowiński M (red.) Język i społeczeństwo. Czytelnik, Warszawa, pp 205–282
Shimray SR (2019) Ways to create awareness on cultural heritage: an overview. Library philosophy and practice 2577. https://digitalcommons.unl.edu/libphilprac/2577. Accessed 12 Feb 2024
Shutaleva A, Martyushev N, Nikonova Z, Savchenko I, Abramova S, Lubimova V, Novgorodtseva A (2022) Environmental behavior of youth and sustainable development. Sustainability 14:250. https://doi.org/10.3390/su14010250
Simşek GG, ElitokKesici AA (2012) Heritage education for primary school children through drama: the case of Aydın, Turkey. Procedia—social and behavioral sciences, 46 (4th world conference on educational sciences (WCES-2012) 02-05 February 2012 Barcelona, Spain, pp 3817–3824. https://doi.org/10.1016/j.sbspro.2012.06.153

Skorowski H (2006) Współczesne ujęcie regionalizmu. Saeculum Christianum Pismo Historyczno-Społeczne 13(2):157–179
Smith L (2006) The uses of heritage. Routledge, London
Smith A (2011) Learning history through heritage place management in the Pacific Islands. J Pac Hist 46(2):228–235. https://doi.org/10.1080/00223344.2011.607271
Srivastava S (2015) A study of awareness of cultural heritage among the teachers at university level. Univ J Educ Res 3(5):336–344
Suchocka A, Królikowska I (2014) Kreowanie tożsamości kulturowej jako wyzwanie XXI wieku. Colloquium Wydziału Nauk Humanistycznych i Społecznych Kwartalnik 4(214):73–88
Świętek A, Brusik K (2022) Kultura islamu w kształceniu geograficznym w Polsce jako egzemplifikacja koncepcji „geografii ciszy". Annales Universitatis Paedagogicae Cracoviensis Studia Geographica 18:36–57
Throsby D (2001) Economics and culture. Cambridge University Press, Cambridge
Tootell H, Freeman M, Freeman A (2014) Generation Alpha at the intersection of technology, play and motivation. In: 2014 47th Hawaii international conference on system sciences, pp 82–90
Towse R (2022) The cultural economy of heritage. In: Towse R, Rizzo I (eds) The economics of heritage. A study in the political economy of culture in Sicily. E. Elgar, Cheltenham, pp 3–19
Ustrzycki AT (2007) Relacje między Polakami a Ukraińcami w Ukrainie. Na marginesie badań nad tożsamością społeczno-kulturową Polaków w Ukrainie. In: Szweda R (ed) Dylematy tożsamości zbiorowych. KUL, Lublin, p 43
Thompson R (2028) Study on the alpha generation and the reflections of its behavior in the organizational environment. J Res Humanit Soc 6(1):09–19
van Dijk JAGM (2010) Społeczne aspekty nowych mediów. Analiza społeczeństwa sieci. Wydawnictwo Naukowe PWN, Warszawa
Wang B, Dai L, Liao B (2023) System architecture design of a multimedia platform to increase awareness of cultural heritage: a case study of sustainable cultural heritage. Sustainability 15(2504):1–15. https://doi.org/10.3390/su15032504
Wendland M (2011) Konstruktywizm komunikacyjny. Wydawnictwo Naukowe IF UAM, Poznań
Wierzbicki A (2000) Wpływ megatrendów cywilizacji informacyjnej na sytuacje w Polsce w początkach XXI wieku. Telekomunikacja i Techniki Informacyjne 1(2):3–17
Wu H (2011) Constructing culturally relevant pedagogy in Chinese heritage language classrooms: a multiple-case study. https://files.eric.ed.gov/fulltext/ED529916.pdf. Accessed 11 Feb 2024
Zabawa M (2011) The influence of English on international internet slang. In: Witalisz A (ed) Migration, narration, communication: cultural exchanges in a globalised world. Internationaler Verlag der Wissenschaften, Frankfurt am Main, pp 167–181
Ziatdinov R, Cilliers J (2021) Generation Alpha: understanding the next Cohort of University Students. Eur J Contemp Educ 10(3):783–789. https://doi.org/10.13187/ejced.2021.3.783
Ziółkowski M (2000) Przemiany interesów i wartości społeczeństwa polskiego. Teorie, tendencje, interpretacje. Wydawnictwo Fundacji Humaniora, Poznań

Practices of Future Learning Processes

How to Develop Spatial Abilities in Kindergarten

Christina Zisi and Aikaterini Klonari

Abstract The purpose of this study is to create an appropriate teaching intervention that will develop kindergarten pupils' spatial abilities. Despite the fact that spatial abilities are referred to in Greek Kindergarten Curriculum, they are not combined with the provision of appropriate teaching interventions, teaching material, or proper teacher training. In our previous research, we found that teaching so far has not resulted in an effective improvement of spatial abilities. An intervention was planned for this purpose and two large-scale giant maps with dimensions of 3 × 4 m were created, a map of the city of Mytilene with a scale of 1:1000, and a map of Lesvos Island with a scale of 1:20,000. The sample was 56 kindergarten pupils (28 boys and 28 girls) aged 4 to 6 years old. Children's knowledge and abilities were tested using a questionnaire before and after the intervention, with the help of kindergarten teachers who wrote the answers as children of this age do not know how to read and write. The findings show that early spatial intervention in kindergarten leads to a significant improvement in pupils' spatial abilities according to the goals set by the curriculum. The improvement was significant regardless of gender and grade (1st or 2nd year in kindergarten). There was a remarkable improvement in total score and in all questions after the teaching intervention.

Keywords Spatial abilities · Kindergarten pupils · Teaching interventions · Large-scale giant maps

C. Zisi (✉) · A. Klonari
University of the Aegean, Mytilene, Lesvos, Greece
e-mail: chrzisi@gmail.com

A. Klonari
e-mail: aklonari@aegean.gr

© The Author(s), under exclusive license to Springer Nature Switzerland AG 2025
D. Piróg and A. Świętek (eds.), *Future-Ready Geography*, Key Challenges in Geography, https://doi.org/10.1007/978-3-031-81268-2_4

Introduction

The different response of each person to the space that coexists with other people and the space itself creates a multitude of meanings for this space (Akyildiz 2020). Spatial thinking is very important for the survival of beings (humans and animals) and the knowledge they develop as they grow (Simms and Gentner 2019). Spatial awareness begins at birth and develops as the pupils grow, while their spatial abilities also improve with education (Pavlovičová and Švecová 2014).

Spatial thinking and spatial abilities coexist in pupils. Spatial thinking is a broader concept and process that relies on spatial abilities to be functionable (Hegarty 2010). Spatial thinking also uses spatial information and abilities to understand and solve problems, while spatial abilities are the individual's mental abilities to transform, generalize, create it mentally, and in this way understand and recall the spatial relationships between objects (Alkouri 2022).

Spatial abilities according to some researchers are classified into "small-scale" spatial abilities, which focus on the encoding and transformation of objects (e.g., mental rotation and mental bisection), and "large-scale" spatial abilities, which focus on the encoding of large spaces and the environment, such as the ability to move across a large environment and geography (Nazareth et al. 2019; Newcombe 2017; Philbeck et al. 2000). Small-scale spatial abilities include spatial perception, mental rotation, and spatial visualization. This classification is mainly based on the allocentric or egocentric frame of reference during mental transformation (Wang and Carr 2020). The allocentric frame refers to the object's central axis, while the egocentric to the observer's body axis. There is also the third frame, the heterocentric one which refers to the perspective of a third person (Ruby and Decety 2001) and is not used as often.

A different approach based on neuropsychology is to categorize spatial abilities based on the use of intrinsic/extrinsic information about objects and static or dynamic movement. The four categories of spatial abilities (Newcombe 2020; Nazareth et al. 2019; Newcombe 2018; Newcombe and Shipley 2015; Hallowel and Hawes 2015) in detail are:

- Instrict (within object)-static. Encoding objects based on the size and arrangement of their parts
- Instrict (within object)-dynamic. Transformation of the object's spatial coding such as rotation, cross-section, folding, and plastic deformation.
- Extrinsic (between objects)-static. Object's spatial position coding in relation to the frame of reference.
- Extrinsic (between objects)-dynamic. Transforming object interactions as one object (or the viewer) moves, so as to maintain a stable representation during movement and allow perspective taking.

Each of the quadrants formed corresponds to specific spatial abilities, as shown in Fig. 1 (Uttal et al. 2012; Nazareth et al. 2019).

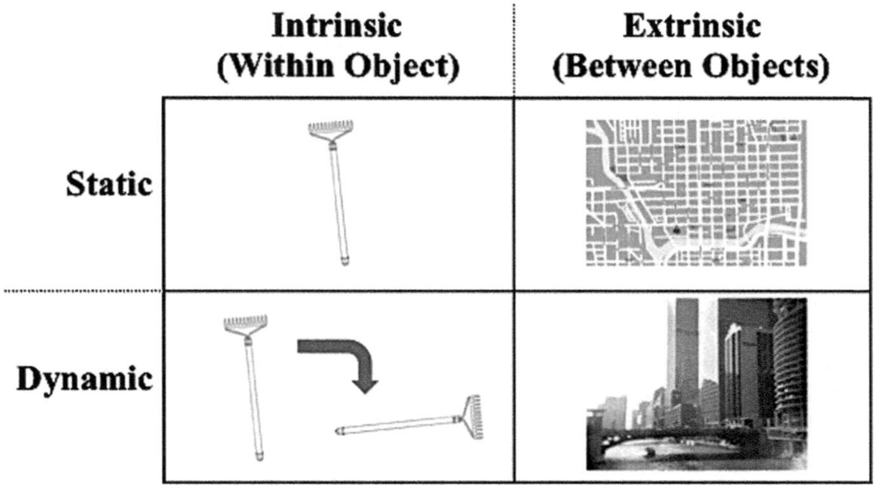

Fig. 1 A 2 × 2 classification of spatial skills (Uttal et al. 2012)

Spatial Abilities in Kindergarten

Kindergarten provides the opportunities to get familiar with the space and acquire spatial abilities. Spatial abilities not only help in the assimilation and utilization of pre-existing knowledge, but also have an important role in the development of new knowledge. Without spatial abilities, creative thinking and innovative creation are incomplete and many activities are unsatisfying (Kell et al. 2013).

Spatial abilities in kindergarten are a stronger predictor even than math skills of children's first-grade achievement (Kotsopoulos et al. 2019).

According to Mohan and Mohan (2013), children aged 3 to 6 are able to:

- Identity and location: recognize places on the map, identify familiar places and landscape features on maps, aerial photographs, and use landmarks to identify places or objects on the map.
- Magnitude: understand the difference in size but confuse the scale with the number of objects that can be contained.
- Relative distance and direction: understand relative distance, e.g., next to, near, etc., and they can use it on the map. They have trouble figuring out which way to hold the map.
- Symbols: Symbols should represent places of the real world and be recognizable images to children. Abstract or unrelated images are not understood.

This lack of interest in teaching them leads to extremely low performance in activities that require spatial abilities. The majority of research on spatial thinking does not deal with the ages from kindergarten to 12 years or with the detailed curriculum for this age. The focus is usually on ages 2–4 or in universities creating a research gap in this developmental period of the individual (Mohan et al. 2015).

Although the curriculum of the Greek kindergarten describes spatial activities, the lack of appropriate teaching intervention for this purpose leads to children who at the end of the school year have very poor spatial abilities. In our previous research, we found that teaching so far has not resulted in an effective improvement of spatial abilities in kindergarten (Zisi et al. 2022). Spatial skills and geography are underrepresented in curriculum and education (Bates et al. 2023; Berson et al. 2023; Gilligan-Lee et al. 2022; Gagnier et al. 2021; Gilligan 2020; Likouri et al. 2017; Klonari and Koutsopoulos 2005). Spatial abilities in contrast to other abilities (linguistic or numerical) are largely neglected (Wang and Carr 2019; Costales et al. 2015; Kell et al. 2013). In recent years, research on spatial abilities has increased, although in schools more emphasis is placed on mathematics and language (Webb et al. 2007). Teachers avoid classroom activities to improve their students' spatial abilities often because of their own poor spatial abilities but also because of their insecurity to engage in spatial activities (Gilligan-Lee et al. 2022; Gagnier et al. 2021; Atit et al. 2020) and having an appropriate teaching intervention that they could use would address the issue (Bates et al. 2023). The improvement of spatial abilities from early school age is particularly important, as individuals who do not receive appropriate training to improve their spatial abilities as children do not have highly developed spatial abilities as adults (Pavlovičová et al. 2022; Duffy et al. 2020; McLaren-Gradinaru et al. 2020; Kavouras et al. 2016).

Early teaching interventions in kindergarten by using maps as spatial educational tool is a way to improve spatial thinking and reduce differences arising from gender and/or socioeconomic status (Newcombe and Frick 2010). Spatial abilities that can be improved with appropriately designed intervention (Rittle-Johnson and Zippert 2019; Klonari 2012: 51) are directly related to individuals' performance in the sciences (Shawky et al. 2020; Sinclair and Bruce 2015).

The Study

This study aims to investigate the suitability of kindergarten to improve children's spatial abilities using an appropriate specially designed teaching intervention for this purpose and the use of large-scale giant maps created specifically for this intervention.

The Method

The study was conducted in Mytilene, on Lesvos Island, Greece, during the school year 2021–2022. In the beginning of the school year, children's spatial abilities were investigated using a questionnaire. The same questionnaire was also used at the end of the school year to find the improvement of their spatial abilities after the implementation of the specially designed teaching intervention with the use of the two large-scale giant maps created especially for this purpose. Due to their age and their inability to read and write, the completion of questionnaire was done by the kindergarten teachers or with the researcher's help. The adults asked the questions and wrote the answers or recorded them. Some questions on how to draw a road, how to create a map, or how to put point symbols on the map were completed by the children themselves. The questionnaire was designed specifically for this research according to the objectives and aims of the Greek kindergarten curriculum. Then, the questions were grouped, categorized, and scored.

The Design and Creation of the Maps

A key tool in our research was the creation by the researchers of two maps, a geomorphological one of the island of Lesvos on a scale of 1:1000, and a political one of the city of Mytilene with a scale of 1:20,000, 3 × 4 m each. The initial data for the maps were provided by the Cartography and Geoinformatics Laboratory of the Department of Geography of the University of the Aegean and updated. A lot of attention was paid to the very good updating of the city of Mytilene political map data as the children should be able to find—during the teaching intervention—on the map their house, the houses of their relatives, or the local shops, etc. The maps were printed on vinyl, so those oversized floor maps to be one-piece, easily wrapped and transported. The really large scale and the giant size of the maps were crucial as they could present a large number of features to be used in the activities, designed to be done on the maps that children can walk on.

The Teaching Intervention

The teaching intervention was designed by the researcher and the activities were based on the spatial skills described in the kindergarten's Curriculum. The teaching intervention was implemented 2 or 3 times a week for 1 or 2 teaching hours, for a period of approximately 2 months. The choice of time was at the discretion of the kindergarten teacher and the only condition was to try to do as many activities as possible than those suggested, but to implement at least one of each spatial ability. The design and implementation of the teaching intervention was based on the Addie model and its five phases (analysis, design, development, implementation, and evaluation).

The planning during the implementation of the teaching intervention was done in such a way as to integrate many spatial abilities, to select the appropriate activities for the age and cognitive level of the children, to integrate problem-solving activities with different levels of difficulty, and then to identify those points where the kindergarten teacher should have focused more (Astleitner 2020).

The teaching intervention was designed to rely on the use of large-scale giant maps created specifically for this purpose. It was designed to be implemented in three stages. In the beginning, introductory activities lasting 1 to 2 days should be carried out. The aim of these activities was to make the children realize that the map is a print on paper of real spaces, with orientation and accuracy, and to understand that it is a print of three-dimensional spaces in a two-dimensional print. At this stage, activities are carried out that will help the children to understand the creation and use of the map. With the help of a compass, they find where north is in their class and place a big arrow, symbol of North with a big B from the word «Βορράς»—north in Greek. They familiarize themselves with the use of the compass, learn the points of the horizon, and realize that the points of the horizon do not change when the position of the observer changes, they are fixed and permanent.

Activities with the Map of the City of Mytilene.

In the next stage, activities were carried out on a large-scale, giant map of the city of Mytilene with a dimension of 3 × 4 m, where children can locate well-known points of interest, their neighborhood, school, park, etc. (Davis and Hyun 2005).

Children realize that a map is a representation of the world it describes—visualizes and is a means of conveying information in a graphic symbolic way. In this map, children learn about map features: title, legend, orientation (B and/or arrow), and learn to "read" the map. They place the map in the correct orientation. The teacher explains that this is what the city of Mytilene looks like if someone sees it from above. He/She explains to the students that the squares correspond to buildings and what the streets and the sea look like. Mytilene is big but with this map, they can play with and on the city. The children lie head to heel in front of each other, form the outline of the city, and count at the end how many children it took to form the outline. They can play walking on "land" or "swimming" in the sea.

The children "read" with the help of the kindergarten teacher the legend and look for known public buildings on the map such as schools, churches, kindergartens, or buildings where their parents work, find their houses or the houses of their acquaintances. They learn to use the map in problem-solving activities such as "I want to go visit my dad at the police station where should I go?" "I'm at school and I want to go to the hospital, but first I have to pick up my brother from the nearby kindergarten, how am I going to go?" They measure distances between various points on the map such as buildings or parks and find that the route they will follow is not a straight line. They circle neighborhoods with wreaths and describe what they contain (e.g., park, school, houses, stadium, etc.) and what they can do there (e.g., play soccer, visit their grandmother, or get on the boat). They follow their road home and understand the need for landmarks so they don't get lost. They see which children live next to them and verbally describe the distance or route they will follow to meet each other. They can drive miniature bikes or cars on the roads. They read from tags the names of the streets in a certain area and try to find them on the map. They indicate by gesturing or walking on the map the desired answer each time. Children are encouraged to use spatial concepts and words correctly. The letters on this map are all capitals.

Activities with the Map of Lesvos Island.

Activities using the large-scale giant map 3X4 meter dimensions of the island of Lesvos. On this map, they can recognize their villages or different places they visited. Activities are done to reinforce the use of the legend, title, and orientation. Children read the title and placed the map in the correct orientation by observing the north symbol high up on the classroom wall. They observe the differences in the relief of the island and explain terms such as mountain, river, bay, altitude, and their color representation using the map legend. Characteristic points such as airports, ports, salt marshes, and their symbols on the map are presented. The teacher explains to the children that the map is the creation of the cartographer and that it is always necessary to read the legend despite the usual conventions that govern a map.

Children move around the map and use cars to travel to villages and remote places. They use the map to navigate by following directions (e.g., one step left, then 2 steps right), solve problems, and make decisions (if I'm in this village, which port will I get to the fastest? I want to buy salt, where shall I go? if I want to go up a mountain to camp where shall I go?). They use symbol cards (Palmer 1994: 16) to show where various points of interest are located, to identify them and then to solve various spatial problems. They make assumptions about distances and lengths of paths and confirm them experimentally. They measure distances in various ways, their hands, feet, tapes, etc. and understand that the distance measured in a straight line on the map is a completely different thing from the distance they will travel by car. They compare distances and routes by measuring roads. They discuss why the roads are not straight and have bends making them considerably longer. They are invited to think about the professions that people do in the different parts of the island and how this depends on the relief and the different features of the map.

They play games where they are blindfolded (wearing a mask) and follow clear orientation instructions to reach their goal and win the prize (a piece of candy), using terms like front, right, left, etc. In groups, they follow orientation instructions to solve the spatial problem given to them using the terms east, north, etc. To help themselves, they look at the north symbol high up on the respective classroom wall. The previous activities are repeated and the children have to reach the same point starting from another point on the map and they have to decide and verify if the routes to reach the same point will be the same or will change and how.

They demarcate points with hoops or ribbons and describe what is in the enclosed space e.g., villages, river, port, salt pans, etc. They choose points of interest and describe what they will do there e.g., they will fish, how they can go (e.g., by car, by boat), if it is near or far, if there are other points of interest there (e.g., a well-known village or a beach for swimming).

In the final stage, children can see the two maps (the map of the island of Lesvos and the map of the city of Mytilene—the capital of the island) side by side to understand the importance of scale in map design and how scale makes a difference to the appearance of the map. The children place the two maps next to each other on the floor and notice the difference in their scale. The one large map of the city of Mytilini is only the red mark on the map of the island of Lesvos. By observing the two maps side by side at the same time, children can clearly see the difference in the two maps at different scales and at the same time understand how large areas are “shrunk” to be depicted on the map.

The Sample

The research sample was 56 participants from 3 kindergarten’s classes.

In every class in Greek kindergarten coexist “pronipia” (on December 31 of this school year reach the age of four (4) years old) completed the 1st year of study in kindergarten and “nipia” (children who on December 31 of this school year reach the age of five (5) years old), completed the 2nd year of study in kindergarten, who attend the same curriculum. Pupils are divided into classes in such a way that each class has a similar number of children for each age group and gender.

Of the 56 pupils, 28 were boys of which 15 were nipia and 13 pronipia. Girls were also 28 (28 of which 13 were nipia and 15 pronipia). In total, there were 28 nipia and 28 pronipia.

Results

The comparison of the mean score of all children and the performance of boys and girls in the beginning (before the teaching intervention) and at the end of the school year (after the teaching intervention) are presented in Table 1.

Comparing the mean value in the beginning and at the end of the school year, the results are impressive and show that: Map recognition mean from 1.82 became 7.95, (max 8), Map usefulness from 0.16 to 2.95, (max 3), Map title from 0.23 to 2.95, (max 3), Map symbols from 1.52 became 7.95, (max 8), Map elements from 0.43 became 3.93, (max 4), Spatial problem solving went from 1.37 to 4.78, (max 5), Map knowledge mean from 1.16 became 3.91, (max 4), Map scale from 1.91 to 2.95, (with a max of 3), Map legend from 0.25 to 4.00, (with a max of 4), Orientation from 0.57 became 3.82, (max 4), Distance perception from 1.13 became 3.77, (max 4), Place identification from 0.36 to 0.93, (with a max of 1), Routes from 0.73 became 2.60 (max 3), Bird eye view went from 0.91 to 2.41, (max 3), Creation of a simple map (way home) from 1.42 became 2.91, (max 3). Total from 13.97 became 57.79, (max 60).

The comparison of the boys’ mean score in the beginning and at the end of the school year in each grouped question and the total score shows that: Map recognition mean from 2.07 became 7.89 (max 8), Map usefulness from 0.18 to 2.96, (max 3), Map title from 0.25 became 2.93 (max 3). Map symbols from 1.25 became 7.96, (max 8), Map element from 0.36 became 3.93, (max 4), Spatial problem solving mean from 1.66 became 4.71, (max 5), Map knowledge mean from 1.39 to 3.86, (max of 4), Map scale from 1.96 to 2.92 (max 3), Map legend from 0.29 to 4.00,

Table 1 Mean score on all children, boys, and girls at the beginning and at the end of the school year

Questions	Pupils (N:56) score						
	Mean pre	Mean post	Mean pre boys	Mean pre girls	Mean post boys	Mean post girls	Max
Map recognition	1.82	7.95	2.07	1.57	7.89	8.00	8.00
Map usefulness	0.16	2.95	0.18	0.14	2.96	2.92	3.00
Map title	0.23	2.95	0.25	0.21	2.93	2.96	3.00
Map symbols	1.52	7.95	1.25	1.79	7.96	7.93	8.00
Map elements	0.43	3.93	0.36	0.50	3.93	3.93	4.00
Problem solving	1.37	4.78	1.66	1.07	4.71	4.83	5.00
Map knowledge	1.16	3.91	1.39	0.93	3.86	3.96	4.00
Scale	1.91	2.95	1.96	1.86	2.92	2.96	3.00
Legend	0.25	4.00	0.29	0.21	4.00	4.00	4.00
Orientation	0.57	3.82	0.50	0.64	3.86	3.79	4.00
Distance perception	1.13	3.77	1.39	0.86	3.82	3.71	4.00
Place identification	0.36	0.93	0.32	0.39	1.00	0.86	1.00
Routes	0.73	2.60	1.02	0.45	2.63	2.57	3.00
Bird eye view	0.91	2.41	0.68	1.14	2.75	2.07	3.00
Creation of a simple map (way home)	1.42	2.91	1.50	1.36	2.89	2.92	3.00
Total performance with all questions answered	13.97	57.79	14.82	13.13	58.13	57.45	60.00

(max 4), Orientation from 0.50 became 3.86, (max 4), Distance perception from 1.39 became 3.82, (max 4). Place identification from 0.32 became 1.00 (max 1), Routes from 1.50 became 2.63 (max 3), Bird eye view mean from 0.68 to 2.75, (max 3), Creation of a simple map (way home) from 1.50 became 2.89 (max 3). The total score mean with all questions was answered from 14.82 became 58.13 (max 60).

When we compare the girls' mean score in the beginning and at the end of the school year, we can observe the impressive improvement in all the questions. Their

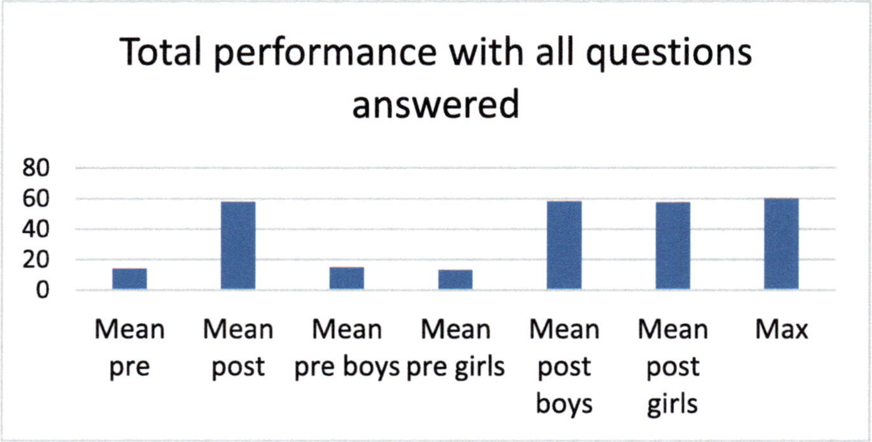

Fig. 2 Total performance of all the children in the beginning and at the end of the school year and the max score

mean score at the end of the school year approaches the maximum possible performance, analytically: Map recognition from 1.57 became 8.00, (max 8), Map usefulness from 0.14 to 2.92, (max 3), Map title from 0.21 to 2.96, (max 3), Map symbols from 1.79 became 7.93 (max 8), Map elements from 0.50 became 3.93, (max 4), Spatial problem solving went from 1.07 to 4.83, (max 5), Map knowledge from 0.93 to 3.96, (max 4), Map scale from 1.86 raise to 2.96, (max 3), Map legend from 0.21 to 4.00, (max 4), Orientation from 0.64 became 3.79, (max 4), Distance perception from 0.86 became 3.71, (max 4). Place identification from 0.39 to 0.86, (max 1). Routes from 0.45 became 2.57, (max 3), Bird eye view from 1.14 to 2.07, (max 3), Creation of a simple map (way home) from 1.36 became 2.92, (max 3). The total score from 13.13 became 57.45 (max 60) (Figs. 2 and 3).

It is obvious that the performance of the children in each grouped question had an impressive increase after the implementation of the teaching intervention. The question about the legend had the greatest improvement as mean score after the implementation of the teaching intervention was 4 (the maximum possible) for all the children and for boys and girls. The girls on the question about map recognition at the end of the school year had a mean of 8 the maximum possible while at the beginning of the school year, the boys had a slightly better mean of 2.07 compared to the girls' 1.57. It appears that girls benefited from the teaching intervention more than boys (mean 7.89) in this particular question. In the place identification question, girls performed slightly better (mean 0.39) than boys (mean 0.32) at the beginning of the school year, but at the end of the school year boys achieved the maximum score (mean 1.00 versus 0.86 for girls).

In the questions: map title, problem solving, map knowledge, scale, and creation of a simple map, the boys had a better score (mean) at the beginning of the school year (and before the teaching intervention) but the girls had a better score (mean) at the end of the school year. In the questions: map symbols, orientation, and bird

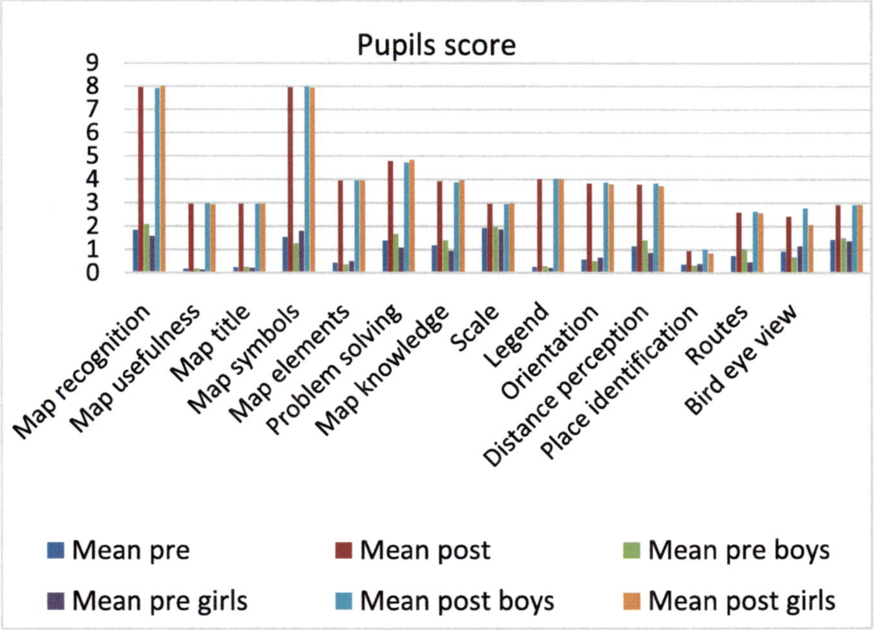

Fig. 3 Mean score of all the children, boys and girls, in the beginning and at the end of the school year

eye view, the girls had a better performance (mean score) at the beginning of the year (and before the teaching intervention) but the boys had a higher performance (mean score) at the end of the school year. In the question about map elements, girls had a slightly better performance at the beginning of the school year but at the end of the school year and after the teaching intervention mean score was exactly the same for boys and girls. In the overall performance initially boys, girls, and all the children had similar mean scores with boys performing better (mean 14.82 and 13.13 for the girls). In the end of the school year, boys have the best performance (mean 58.13) with girls' performance being very close (57.45).

In the beginning of the school year and before the implementation of the proposed teaching intervention, the performance of the children both in total and in the subsets of boys and girls is very low and their overall performance (mean) is less than a quarter of the maximum score. At the end of the school year and after the implementation of the teaching intervention, the mean score of all the pupils, boys, and girls, had a great improvement, which approached the best possible score.

The comparison of mean scores between the two different age groups (classes) is presented in the following Table 2.

The difference in mean score of nipia is observed both in grouped questions and in overall performance. The impressive improvement can be seen in the detailed presentation of the mean score of the nipia (elder kids) in the beginning and at the end

Table 2 Mean performance of nipia (1st year in kindergarten) and pronipia (2nd year in kindergarten) in the beginning and at the end of the school year

Questions	Pupils (*N* nipia:28, pronipia: 28) score				
	Mean pre nipia	Mean pre pronipia	Mean post nipia	Mean post pronipia	Max
Map recognition	2.29	1.36	7.96	7.92	8.00
Map usefulness	0.11	0.21	2.93	2.96	3.00
Map title	0.29	0.18	2.93	2.96	3.00
Map symbols	2.00	1.04	8.00	7.89	8.00
Map elements	0.29	0.57	4.00	3.86	4.00
Problem solving	1.46	1.27	4.96	4.59	5.00
Map knowledge	1.57	0.75	4.00	3.82	4.00
Scale	2.39	1.43	3.00	2.89	3.00
Legend	0.29	0.21	4.00	4.00	4.00
Orientation	0.54	0.61	3.89	3.75	4.00
Distance perception	1.29	0.96	3.82	3.71	4.00
Place identification	0.46	0.25	0.96	0.89	1.00
Routes	1.16	0.30	2.80	2.39	3.00
Bird eye view	1.07	0.75	2.64	2.18	3.00
Creation of a simple map (way home)	1.75	1.11	2.93	2.89	3.00
Total performance with all questions answered	16.95	11.00	58.84	56.73	60.00

of the school year: Map recognition mean from 2.29 became 7.96, (max of 8), Map usefulness from 0.11 to 2.93, (max of 3), Map title from 0.29 to 2.93, (max 3), Map symbols from 2.00 became 8.00 (max 8), Map elements from 0.29 became 4.00, (max 4), Spatial problem solving went from 1.46 to 4.96, (max 5), Map knowledge from 1.57 became 4.00, (max 4), Map scale from 2.39 to 3.00 (max 3), Map legend from 0.29 to 4.00, (max 4), Orientation from 0.54 became 3.89, (with max 4), Distance perception from 1.29 became 3.82, (max 4), Place identification from 0.46 to 0.96, (max 1), Routes from 1.16 became 2.80, (max 3), Bird eye view went from 1.07 to 2.64, (max 3), Creation of a simple map (way home) from 1.75 became 2.93, (max 3). Total performance with all questions answered from 16.95 became 58.84, (max 60).

The comparison of the mean score of the pronipia (preschoolers-younger pupils) in the beginning and at the end of the school year in each grouped question and the total performance with all questions answered shows that: Map recognition mean from 1.36 became 7.92, (max 8), Map usefulness from 0.21 to 2.96, (max 3), Map title from 0.18 to 2.96 (max 3), Map symbols mean from 1.04 became 7.89, (max 8), Map elements from 0.57 became 3.86 (max 4), Spatial problem solving went from

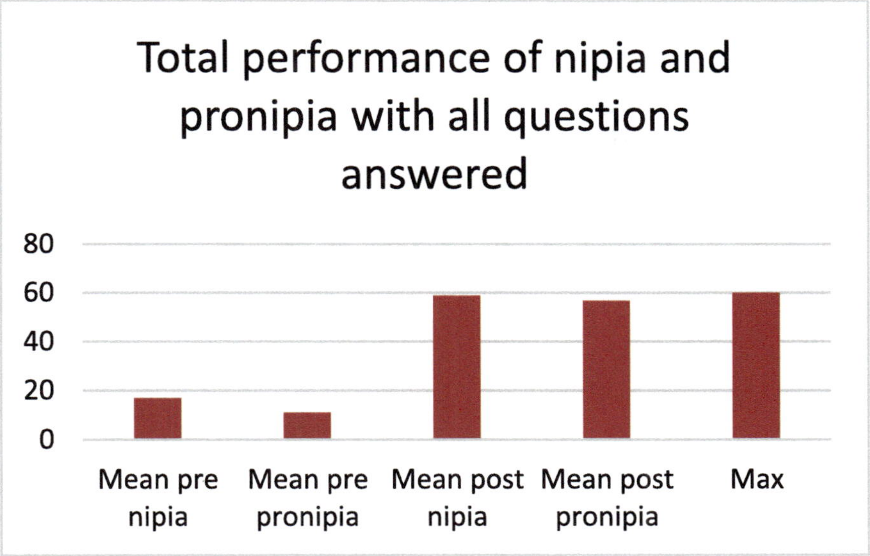

Fig. 4 Total performance of nipia and pronipia in the beginning and at the end of the school year and the max score

1.27 to 4.59, (max 5), Map knowledge from 0.75 to 3.82, (max 4). Map scale from 1.43 to 2.89 (max 3). Map legend from 0.21 to 4.00 (max 4). Orientation from 0.61 became 3.89 (max 4), Distance perception from 1.29 became 3.75 (max 4). Place identification from 0.25 to 0.86 (with a max of 1). Routes mean from 0.30 became 2.39 (max 3). Map view from 0.75 to 2.18 (max 3). Creation of a simple map (way home) from 1.11 became 2.89 (max 3). The total score mean from 11.00 became 56.73 (max 60) (Figs. 4 and 5)@@.

In the question about legend in the beginning of the school year and before the implementation of the proposed teaching intervention, both classes had a very low performance (mean: pronipia 0.29 and nipia 0.21), but at the end of the school year, they had a mean of 4 the maximum possible achievement both. It is observed that children can understand the use of legend and can use it when taught through appropriate learning activities regardless of their class and age (as nipia are 1 year elder than pronipia).

The difference in the beginning of the school year in performance (mean scores) between nipia (2nd year of kindergarten) and pronipia (1st year of kindergarten) is small in relation to the maximum possible performance. Nipia performs a little better than a quarter of their maximum possible score and pronipia a little less. There are questions in which pronipia—who are just starting kindergarten and haven't had time to develop their spatial abilities—perform better (mean score) than nipia who have already attended a year of kindergarten. The questions that pronipia perform better in the beginning of the school year are map usefulness (mean 0.21 and 0.11), map elements (mean 0.57 and 0.29), and orientation (mean 0.61 and 0.54). The fact that

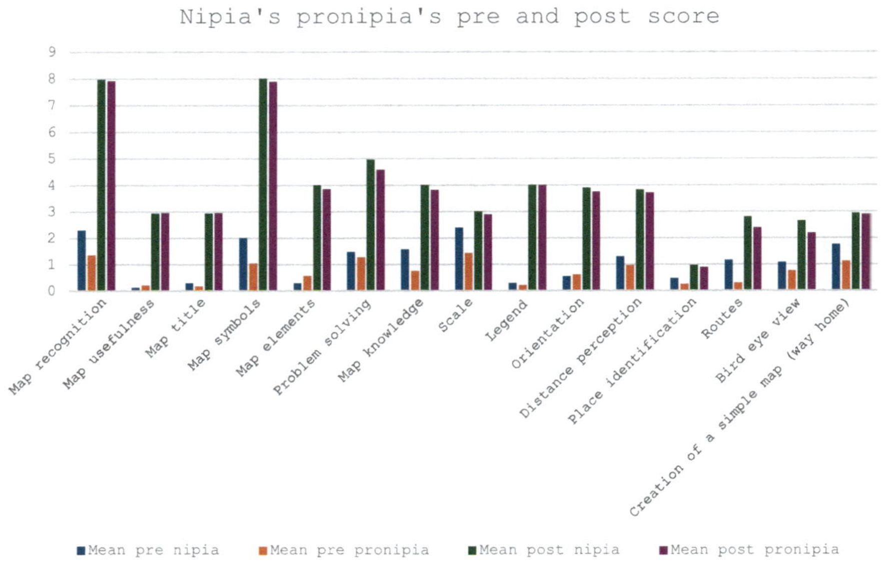

Fig. 5 Nipia's (1st year in kindergarten) and pronipia's (2nd year in kindergarten) score, at the beginning and at the end of the school year

the younger children (pronipia) perform better even though they all have extremely low performance indicates that these three spatial abilities were never taught to the children who had already attended one year of kindergarten or were taught in an ineffective way and the children did not improve their abilities. At the end of the school year in the questions about the usefulness of the map and the title of the map, pronipia had a slightly better performance (mean 2.96) than nipia (mean 2.93) with a maximum score of 3 while nipia had a slightly higher mean score on most questions although both groups had scores approaching the maximum score.

Conclusion

The findings of this research show the great improvement of children's spatial abilities and the effectiveness of the proposed teaching intervention using the two specially designed and created large-scale giant maps. It has been shown in this study that by using hands on and playful activities and encouraging children to move and use gestures according to our teaching intervention and agreeing with previous research (Lee-Cultura and Giannakos 2020; Galati et al. 2018; Wassenburg et al. 2018), they can make dramatic improvements in their spatial abilities. Our research agrees with previous research (Gifford et al. 2022; Alkouri 2022; Zisi et al. 2021) that reached the same conclusion. All children improved their spatial abilities on all grouped questions, and there were no questions (corresponding to spatial abilities) with little or no improvement. Spatial abilities can be improved with appropriate teaching interventions thanks to their plasticity (Zhu et al. 2023; Xu et al. 2023; Yang et al. 2020). At the end of the school year, the children had more than doubled their scores, and each group's mean performance was very close to the maximum performance on each grouped question.

Comparing the scoreboards at the beginning and at the end of the school year, the impressive improvement in children's spatial abilities is evident regardless of class and age (nipia, older children, 2nd year of kindergarten or pronipia, children one year younger, 1st year of kindergarten) and gender. Girls increased their performance more (overtaking boys who had better performance initially) on the questions: map title, problem solving, map knowledge, scale, and creation of a simple map. Boys increased their performance (overtaking girls who had better performance initially) on the questions map symbols, orientation, bird eye view. Boys and girls had an excellent improvement in their overall score but slightly better improvement in different abilities. Although boys had a slightly better performance (mean 58.13 versus 57.45 for girls), it is evident that the proposed teaching intervention was effective for both boys and girls. When boys and girls have the same access to spatial activities, there are no gender differences in their performance as it was found in previous research (Zhang et al. 2020).

There were three questions about map usefulness, map elements, and orientation in which pronipia in the beginning of the school year performed slightly better than nipia who had already attended a year of kindergarten. This suggests that these spatial abilities were either not taught at all or not effectively to the nipia (elder kids, 2nd year in kindergarten) when they were in their first year of kindergarten. At the end of the school year and after the implementation of the teaching intervention, nipia had better score in almost all questions but there were two questions, one on the usefulness of the map and the other on the title of the map, on which pronipia performed slightly better than nipia, although the performance of both groups is close to the maximum possible. It is observed that in these questions it was the proposed teaching intervention with the use of the two large-scale giant maps that developed the corresponding spatial abilities while the small age difference between the two classes (one year) was not a determining factor.

Despite the extremely low pupil's performance (mean score) in the beginning of the school year, with the pupils' groups, overall, boys, girls, and pronipia achieving less than a quarter of the desired maximum performance and nipia just over a quarter (mean score:16.95, max score: 60) at the end of the school year all pupils' groups had a performance (mean score) that was close to the maximum possible. The results were excellent (with very small differences) regardless of age and gender and according to the requirements and suggestions of the Kindergarten curriculum. The proposed teaching intervention with the use of the two large-scale giant maps designed to be tailored to kindergarten children, was effective for all children and helped them dramatically improve their spatial abilities. Teaching spatial abilities has been found to be neglected (Bates et al. 2023; Berson et al. 2023; Zisi et al. 2022; Gilligan-Lee et al. 2022; Gagnier et al. 2021; Gilligan 2020; Wang and Carr 2019), and the proposed teaching intervention may be a response to the lack of appropriate teaching interventions to improve spatial abilities.

For kindergarten to be appropriate for the development of spatial abilities, age-appropriate teaching interventions using educational materials specifically designed for this purpose are required (Zisi et al. 2022). The results confirm the effectiveness of the proposed teaching intervention with the use of large-scale giant maps in kindergarten children and at the same time highlight the kindergarten as suitable for developing children's spatial abilities with the proposed intervention (or other specially designed teaching intervention with the use of appropriate teaching material creating or collecting for this purpose). The proposed teaching intervention is an answer to the question of how to develop spatial abilities in kindergarten.

The small sample and the conduct of the research only in the city of Mytilene on the island of Lesvos does not allow the generalization of the findings. Although they are very encouraging, further research with a larger sample size and in more diverse areas is needed. It is important to find or confirm teaching interventions that will help develop children's spatial abilities from kindergarten age.

References

Akyildiz NA (2020) The concept of space from the space-individual relationship perspective. Int J Innov Sci Res Technol 5(2). ISSN No: 2456–2165

Alkouri Z (2022) Developing spatial abilities in young children: implications for early childhood education. Cogent Educ 9(1). https://doi.org/10.1080/2331186X.2022.2083471

Astleitner H (2020) Alternative theoretical frameworks for educational interventions. In: Astleitner H (eds) Intervention research in educational practice. Waxmann, Munster-New York, pp 19–38

Atit K, Uttal DH, Stieff M (2020) Situating space: using a discipline-focused lens to examine spatial thinking skills. Cogn Res: Principles Impl 5(19):1–16. https://doi.org/10.1186/s41235-020-00210-z

Bates KE, Williams AY, Gilligan-Lee KA, Gripton C, Farran EK (2023) Practitioner's perspectives on spatial reasoning in educational practice from birth to 7 years. Br J Educ Psychol 93(2):571–590. https://doi.org/10.1111/bjep.12579

Berson IR, Berson MJ, McKinnon C, Aradhya D, Alyaeesh M, Luo W, Shapiro BR (2023) An exploration of robot programming as a foundation for spatial reasoning and computational thinking in preschoolers' guided play. Early Childhood Res Q 65:57–67. https://doi.org/10.1016/j.ecresq.2023.05.015

Costales A, Abad C, Odean R, Pruden SM (2015) Spatial activities and manipulatives for early education classrooms. In: Scarlett WG (ed) The SAGE Encyclopedia of classroom management. SAGE Publications, New York pp 768–771. ISBN 978-1-4522-4139-5

Davis AG, Hyun E (2005) A study of kindergarten children's spatial representation in a mapping project. Math Educ Res J 17(1):73–100. https://doi.org/10.1007/BF03217410

Duffy G, Sorby S, Bowe B (2020) An investigation of the role of spatial ability in representing and solving word problems among engineering students. J Eng Educ 109(3):424–442. https://doi.org/10.1002/jee.20349

Gagnier KM, Holochwost SJ, Fisher KR (2021) Spatial thinking in science, technology, engineering, and mathematics: elementary teachers' beliefs, perceptions, and self-efficacy. J Res Sci Teach 59(1):95–126. https://doi.org/10.1002/tea.21722

Galati A, Weisberg SM, Newcombe NS, Avraamides MN (2018) When gestures show us the way: co-thought gestures selectively facilitate navigation and spatial memory. Spat Cogn Comput 18(1):1–30. https://doi.org/10.1080/13875868.2017.1332064

Gifford S, Gripton C, Williams HJ, Lancaster A, Bates KE, Williams AY, Gilligan-Lee K, Borthwick A, Farran EK (2022) Spatial reasoning in early childhood. https://doi.org/10.31234/osf.io/jnwpu

Gilligan KA (2020) Make space: the importance of spatial thinking for learning mathematics. Front Young Minds 8(50). https://doi.org/10.3389/frym.2020.00050

Gilligan-Lee KA, Hawes ZCK, Mix KS (2022) Spatial thinking as the missing piece in mathematics curricula. NPJ Sci Learn 7(10). https://doi.org/10.1038/s41539-022-00128-9

Hallowell D, Hawes Z (2015) The development of spatial reasoning in young children. In: Brent D (ed) Spatial reasoning in the early years. Principles, assertions, and speculations. Routledge, New York

Hegarty M (2010) Components of spatial intelligence. Psychol Learn Motivation 52:265–297. ISSN 0079–7421, https://doi.org/10.1016/S0079-7421(10)52007-3

Kavouras M, Darra A, Kokla M, Kontaxakis S, Panopoulos G, Tomaki E (2016) Geographic information science. Integrated approach and specific issues. Athens: Greek academic e-books and AIDS. Retrieved from File:///C:/Users/admin/Downloads/00_master_document.pdf

Kell HJ, Lubinski D (2013) Spatial ability: a neglected talent in educational and occupational settings. Roeper Rev 35:219–230, ISSN: 0278–3193 print/1940–865X online. https://doi.org/10.1080/02783193.2013.829896

Kell HJ, Lubinski D, Steiger JH (2013) Creativity and technical innovation spatial ability's unique role. Psychol Sci 24(9):1831–1836. https://doi.org/10.1177/0956797613478615

Klonari A (2012) Primary school pupils' ability to use aerial photographs and maps in the subject of geography. Eur J Geogr 3(2):42–52

Klonari A, Koutsopoulos K (2005) Primary and secondary educators' attitudes on school geography. In: Donert K, Charzynski P(eds) Changing horizons in geography education. Herodot Network, Torun, Poland, pp 151–155

Kotsopoulos D, Makosz S, Zambrzycka J, Dickson BA (2019) Individual differences in young children's visual-spatial abilities. Early Child Dev Care 191(14):2246–3225. https://doi.org/10.1080/03004430.2019.1699918

Lee-Cultura SL, Giannakos M (2020) Embodied interaction and spatial skills: a systematic review of empirical studies. Interact Comput 32(4):331–366. https://doi.org/10.1093/iwcomp/iwaa023

Likouri A, Klonari A, Flouris G (2017) Relationship of pupils' spatial perception and ability with their performance in geography. RIGEO 7(2):154–170. Retrieved from chrome-extension: https://dergipark.org.tr/tr/download/article-file/590188

McLaren-Gradinaru M, Burles F, Dhillon I, Retsinas A, Umiltà A, Hannah J, Dolhan K, Iaria G (2020) A novel training program to improve human spatial orientation: preliminary findings. Front Hum Neurosci 14(5). https://doi.org/10.3389/fnhum.2020.00005

Mohan AL, Mohan L (2013) Spatial thinking about maps: Development of concepts and skills across the early school years. Report prepared for National Geographic Education Programs

Mohan L, Mohan A, Uttal D (2015) Research on thinking and learning with maps and geospatial technologies. In: Solem M, Huynh NT, Boehm R (eds) Learning progressions for maps, geospatial technology, and spatial thinking: a research handbook. Association of American Geographers, Washington, DC, pp 9–22

Nazareth A, Newcombe NS, Shipley TF, Velazquez M, Weisberg SM (2019) Beyond small-scale spatial skills: navigation skills and geoscience education. Cogn Res: Principles Impl 4(17). https://doi.org/10.1186/s41235-019-0167-2

Newcombe N (2017) Harnessing spatial thinking to support stem learning. OECD Education Working Papers, 161, OECD Publishing, Paris. https://doi.org/10.1787/7d5dcae6-en

Newcombe NS (2020) The puzzle of spatial sex differences: current status and prerequisites to solutions. Child Dev Perspec 14(4):251–257. https://doi.org/10.1111/cdep.12389

Newcombe NS, Frick A (2010) Early education for spatial intelligence: why, what, and how. Mind Brain Educ 4(3):102–111. https://doi.org/10.1111/j.1751-228X.2010.01089.x

Newcombe NS, Shipley TF (2015) Thinking about spatial thinking: new typology, new assessments. In: Gero J (ed) Studying visual and spatial reasoning for design creativity. Springer, Dordrecht. https://doi.org/10.1007/978-94-017-9297-4_10
Pavlovičová G, Švecová V (2014) The development of spatial skills through discovering in the geometrical education at primary school. In: Proceedings of the 5th world conference on learning, teaching and educational leadership, WCLTA 2014. https://doi.org/10.1016/j.sbspro.2015.04.189
Pavlovičová G, Bočková V, Laššová K (2022) Spatial ability and geometric thinking of the students of teacher training for primary education. TEM J 11(1):388–395. https://doi.org/10.18421/TEM111-49
Philbeck JW, Behrmann M, Black SE, Ebert P (2000) Intact spatial updating during locomotion after right posterior parietal lesions. Neuropsychologia 38(7):950–963. https://doi.org/10.1016/s0028-3932(99)00156-6
Rittle-Johnson B, Zippert E (2019) The roles of patterning and spatial skills in early mathematics development. Early Childhood Res Q 46(1):166–178. https://doi.org/10.1016/j.ecresq.2018.03.006
Ruby P, Decety J (2001) Effect of subjective perspective taking during simulation of action: a PET investigation of agency. Nat Neurosci 4(5):1–5
Shawky A, Elbiblawy E, Maresch G (2020) Spatial ability differences between students with a math learning disability and their other normal colleagues. J Human Appl Soc Sci 3(3):182–198. https://doi.org/10.1108/JHASS-01-2020-0016
Simms NK, Gentner D (2019) Finding the middle: Spatial language and spatial reasoning. Cogn Dev 50:177–194. https://doi.org/10.1016/j.cogdev.2019.04.002
Sinclair N, Bruce CD (2015) New opportunities in geometry education at the primary school. ZDM Mathematics Education 47:319–329. https://doi.org/10.1007/s11858-015-0693-4
Uttal DH, Meadow NG, Tipton E, Hand LL, Alden AR, Warren C, Newcombe NS (2012) The malleability of spatial skills: a meta-analysis of training studies. Psychol Bull 139(2):352–402. https://doi.org/10.1037/a0028446
Wang L, Carr M (2019) Gender, working memory, strategy use, and spatial ability. N Am J Psychol 21(3):601–618
Wang L, Carr M (2020) Understanding the relationship between spatial ability and mathematics achievement through the lens of information processing theory. North Am J Psychol 22(2):173–192. https://doi.org/10.13140/RG.2.2.20223.15524
Wassenburg SI, De Koning BB, Van Der Schoot M (2018) In which direction to move? Facilitative and interference effects of gestures on problem solver's thinking. J Cogn Psychol 30(3):307–313. https://doi.org/10.1080/20445911.2018.1432628
Webb RM, Lubinski D, Benbow CP (2007) Spatial ability: a neglected dimension in talent searches for intellectually precocious youth. J Educ Psychol 99(2):397–420. ISSN: 1939–2176. https://doi.org/10.1037/0022-0663.99.2.397
Xu S, Song Y, Liu J (2023) The development of spatial cognition and its malleability assessed in mass population via a mobile game. Psychol Sci 34(3):345–357. https://doi.org/10.1177/09567976221137313
Yang W, Liu H, Chen N, Xu P, Lin X (2020) Is early spatial skills training effective? a meta-analysis. Front Psychol 11(1938). https://doi.org/10.3389/fpsyg.2020.01938
Zhang X, Chen C, Yang T, Xu X (2020) Spatial skills associated with block-building complexity in preschoolers. Front Psychol 11(563493). https://doi.org/10.1016/j.cedpsych.2017.09.004
Zhu C, Leung COY, Lagoudaki E, Velho M, Segura-Caballero N, Jolles D, Duffy G, Maresch G, Pagkratidou M, Klapwijk R (2023) Fostering spatial ability development in and for authentic STEM learning. Front Educ 8. https://doi.org/10.3389/feduc.2023.1138607

Zisi C, Klonari A, Soulakellis N, Tataris G (2021) Introducing geography and reading map skills to kindergarten children by using large-scale giant maps. Int J Educ 9(4):1–17. https://doi.org/10.5121/ije2021.9401

Zisi C, Klonari A, Lambrinos N (2022) Investigating kindergartners' spatial abilities. Are they the best they can be? Eur J Geogr 13(3):1–18. https://doi.org/10.48088/ejg.c.zis.13.3.001.018

How Can Scientific Literacy Be Enhanced in Geography Education? Two Ways Forward: Geoboxes and Digital Experiments

Nina Scholten, Natalie Bienert, and Rainer Mehren

Abstract In addressing the pressing global challenges of the future such as climate change or biodiversity loss, geography education stands out for its unique strength in offering an integrated perspective of (natural) science and social science. Despite this strength, there is a noticeable absence of the science perspective in geography lessons in Germany. Inquiry-based science education in geography classrooms is rare, which poses a challenge to achieve a holistic geographical perspective and scientific literacy. To address this need, two approaches to implement a stronger science perspective in geography classrooms are presented: Firstly, science kits (Geoboxes): Geoboxes comprise all materials and resources to conduct inquiry-based science education in Geography. For example one of the Geoboxes deals with an experiment about soil erosion. Geoboxes come in a compact box ready-to-use in the classroom. Geoboxes provide teachers with a guided access to science education and additional instructional material. Secondly, digital videos that showcase geographical experiments are introduced (ViExGeo). The open access videos are structured along the scientific algorithm. They come with a teaching guide, which offers different instructional settings to support scientific literacy among geography students. Both approaches are critically discussed.

Keywords Scientific literacy · Science kits · Experiments · Scientific inquiry · Inquiry-based learning

---

The original version of the chapter has been revised. A correction to this chapter can be found at https://doi.org/10.1007/978-3-031-81268-2_13

N. Scholten (✉)
Geography Education, Universität Hildesheim, Hildesheim, Germany
e-mail: nina.scholten@uni-hildesheim.de

N. Bienert · R. Mehren
Geography Education, Universität Münster, Münster, Germany
e-mail: natalie.bienert@uni-muenster.de

R. Mehren
e-mail: rainer.mehren@uni-muenster.de

© The Author(s), under exclusive license to Springer Nature Switzerland AG 2025, corrected publication 2025

D. Piróg and A. Świętek (eds.), *Future-Ready Geography*, Key Challenges in Geography, https://doi.org/10.1007/978-3-031-81268-2_5

Introduction

In addressing the pressing global challenges of the future, such as climate change, food insecurity, energy transition, and biodiversity loss, geography stands out for its unique strength in offering an integrated perspective of natural science and social science. This perspective is crucial because complex challenges or wicked problems demand a multidisciplinary view to search for adequate and sustainable responses. Since a systemic perspective, which integrates human and physical geography, is constitutive for geography, this needs to be represented in the school subject. Additionally, many empirical studies have shown that curricula integrating science into the geography classroom have yielded positive results in terms of students' achievement (Eidietis and Ruthford 2009; Grubbs and Grubbs 2015; Oldakowski and Johnson 2018).

However, the subject in schools only partially lives up to this demand in everyday school life. There are several indicators that the science side of geography is underrepresented in geography instruction (Lethmate 2005; Otto 2003). One empirical indicator entails that scientific inquiry practices find little entry into geography instruction. Experiments are rarely conducted, although there is a high interest among students in experimenting. A study by Hemmer and Hemmer (2010) asked 3741 students in Bavaria to rank 17 methods according to their interest. The study revealed that among students, the most popular method is the experiment. The same list was presented to geography teachers ($n = 39$), who were asked to indicate how frequently they employ each method in their teaching. The results showed that experiments were seldomly used by geography teachers.

Another manifestation of neglect can be found in geographical teaching resources. If scientific education is integrated into geographical teaching resources, such as textbooks, the representation of science education often shows shortcomings and deficiencies. Lethmate (2005, 2006) has repeatedly criticized that in many textbooks any kind of investigation is called an "experiment", even if the activities did not remotely follow an experimental design (Lethmate 2005). In other cases, experiments are described prescriptively, having students follow step-by-step instructions in a cookbook fashion ("First, you do this, then that, then…"). Students work through the step-by-step guide, doing something without any cognitive challenge or scientific reasoning. Often, it seems as if scientific insights concerning geographical issues are simply reported in a text or conveyed via a figure and a task. This way, students are informed about scientific results without active engagement.

Additionally, students' competencies concerning scientific literacy seem rather low. Peter and Sprenger (2014) investigated geography students' experimental competencies, subdivided according to an experimental design: formulating a research question, generating hypotheses, planning the experiment, and analysis of data (Grube 2010), with the help of a questionnaire. Although results vary concerning the different aspects, no aspect achieves a high average level. The authors advocate for more scientific education in geography. In addition, the past and current PISA results emphasize difficulties with scientific literacy among German students (Schwippert et al. 2020).

The limited implementation of science education in geography classrooms seems not restricted to Germany. Other countries face similar challenges (e.g., Caldis and Kleeman 2021; Orion and Libarkin 2023; Tomcikova 2020; Vasileva and Poleganova 2017; Xuan et al. 2019). The subject is at risk of weakening geography's educational potential, the integrative perspective of social and natural science, when science education is not adequately considered in geography classrooms (Vasileva and Poleganova 2017; Xuan et al. 2019). To bridge this gap, the aim of this book chapter is to introduce and discuss two approaches to support science in geography classrooms. Both approaches are easily applicable and adaptable for geography classrooms in other contexts.

For this purpose, understanding the causes for the neglect of science education in geography instruction needs to be analyzed to understand which conditions facilitate or impede science education. Secondly, the overarching aim of science education, scientific literacy, is outlined along with ways to achieve scientific literacy. Building on that, two approaches (Geobox and ViExGeo) are presented. Finally, the two approaches are assessed in relation to the constraints and their contribution to scientific literacy.

(Insufficient) Science in Geography Education

Causes for Neglect of Science in Geography Education

Several factors contribute to the marginal position of science in geography lessons. Empirical studies show that the limited implementation of science teaching is partly due to contextual factors or structural conditions. (Future) Geography teachers report being constrained by various factors, including the limited time frame of the subject, as inquiry-based science teaching requires longer teaching blocks (Höhnle and Schubert 2016; Tomcikova 2020; Vasileva and Poleganova 2017), a packed curriculum that needs to be covered (Höhnle and Schubert 2016), large class sizes (Höhnle and Schubert 2016; Tomcikova 2020), and the unavailability of supplies and scientific facilities (Höhnle and Schubert 2016; Vasileva and Poleganova 2017). This closely echoes findings by Akuma and Callaghan (2019), who investigated science teachers' inhibitions of inquiry-based practical work. Additionally—and perhaps more significantly—many geography teachers feel uncomfortable teaching science (Hof and Hennemann 2013; Otto 2003; Tomcikova 2020). They feel insufficiently prepared or perceive their competencies in and understanding of science teaching as inadequate. For instance, the study by Hof and Hennemann (2013) indicates that novice teachers perceive their acquired competencies in experimenting during the initial teacher education (ITE) as insufficient. Additionally, Höhnle and Schubert (2016) identify among the top 10 hindrances for science education according to geography teachers:

"inadequate preparation for the use of experiments, etc., during initial teacher training" (p. 156) and "lack of teachers' knowledge for the specific integration … in teaching" (p. 156). This is not surprising because even for science teachers understanding what scientific inquiry is and what scientific inquiry looks like in practice remains challenging (Akuma and Callaghan 2019; Schwartz et al. 2023). Besides the self-attributed inability, it seems as if geography teachers, especially those without an additional science subject, seldomly identify as science teachers. They rather think of themselves as social science teachers and, with this reasoning, reject teaching science (Lethmate 2005). In Germany, this attitude is supported by geography's school curriculum, which is housed within social sciences (Lethmate 2005; Otto 2003). This classification is consequential for geography teaching as it "affects the content within the curriculum, the backgrounds of teachers and their professional development opportunities, and also the expectations and subject choices of students" (Selway 2021, 18).

Scientific Literacy from a Geographic Perspective

Scientific Literacy describes the desired general outcome of learning science (Holbrook and Rannikmae 2009; Schneegans and Nair-Bedouelle 2021). Since the term was first proposed in the 1950s, it has encompassed different concepts and interpretations (Xuan et al. 2019). In general, it includes (1) understanding foundational and core science concepts, (2) scientific inquiry practices, and (3) epistemic features and functions of those concepts and practices, which is also called nature of science (NOS) (Schwartz et al. 2023). The three basic components will be outlined in more detail below:

The first component (1) refers to knowledge of facts, concepts, principles, laws, hypotheses, theories, and models of science (Xuan et al. 2019). For instance, the German national standards for Geography education comprise geoscience topics such as earth's fundamental characteristics, earth's position and movement, interactions of the natural spheres of the earth, physical geographical phenomena and structures, spatial physical geographical processes, geographic factors, and simple cycles (German Geographical Society 2014). This includes the appropriate and adequate use of scientific language (Schwartz et al. 2023).

The concept of scientific literacy emphasizes that science is not only a body of knowledge that reflects the current understanding of the world but also (2) a set of practices used to establish, extend, and refine that knowledge (Schwartz et al. 2023). There is widespread agreement that the exploring and discovering process of scientific inquiry is important in the science classroom. Inquiry-based learning reflects the practices of actual scientists. It includes engaging science learners in investigating the natural world as they pose researchable questions and investigate the question with the help of inquiry practices, in addition to formulating evidence-based explanations and justifications of their assertions (Akuma and Calleghan 2019). Three practices are fundamental to scientific inquiry from a geographical point of view:

observation, investigation, and experimentation (Lethmate 2006; Mönter and Otto 2016; Otto and Mönter 2015). Observation refers to the purposeful, systematic, and conscious perception of a geographic phenomenon or process within its context. Observation is guided by hypotheses and theories, thus relating to previous knowledge. This is precisely what distinguishes observation from a random discovery, as discovery involves noticing something that was not previously considered (Mönter and Otto 2016). An investigation is an observation conducted with devices, with which objects or processes are often intervened. For example, when a natural object (e.g., rock, soil, cotton) is mechanically divided, disassembled, or cut, or when it is treated with chemicals. Partly technical devices are also used for this purpose, for example, conducting a soil profile with a soil auger. Investigations serve, among other things, to gain a better understanding of structures (Otto and Mönter 2015). Often investigations involve measurements that quantify and refine observations. When measuring, an unknown quantity (e.g., room temperature) is compared with a predefined, calibrated unit (e.g., °C) using a measuring device (e.g., a thermometer), and the result is expressed as a numerical value, the measurement value (e.g., 19 °C), along with the unit. The purpose of measurements is to obtain a statement about a quantity. Measurement provides objective data. How these data are interpreted afterward is up to humans (e.g., "Is 18 °C and cloudy considered good weather?"). Common students' measurement errors involve, for example, a poor measurement setup (e.g., hanging the thermometer in direct sunlight), or a lack of standardization in measurement repetitions, incorrect reading of the scale, or misunderstanding the unit.

An experiment, ultimately, represents the continuation of observations and investigations under artificially altered and controlled conditions. Experimenting is a central scientific inquiry practice (Tomcikova 2020). Experiments can be characterized by the following features: Experiments serve to identify cause-effect relationships, consequently, answer questions of causality. The aim of an experiment is to derive regularities (= there are exceptions) or laws (= without exceptions). Experiments are conducted to search for proof of a causal relationship between an independent variable (cause) and a dependent variable (effect). The observation in experiments is not random but prepared or planned and set in artificial conditions. The variable under investigation is clearly defined, isolated, and altered (Isolation/Variation) while all other circumstances are kept constant. An experiment always includes a control approach that identifies and varies the variable under investigation (Control observation). If the experiment is repeated under the same conditions, similar results can be obtained (Replicability) (Otto and Mönter 2015). Geography experiments differ in context compared to other subjects; while other subjects focus on explaining cause-and-effect relationships in natural phenomena, in geography, the interrelation of the Earth's system and humankind is important (Naujoks et al. 2021). An experiment usually follows a reasonably clearly defined set of stages, even though the designations and what they encompass may slightly differ among authors (Otto and Mönter 2015; Tomcikova 2020; Vasileva and Poleganova 2017).

These stages support learners in developing rigorous methods. Kruit and colleagues (2018) found that explicit instruction about the practice and epistemic function of investigation design encouraged students to transfer these skills to new scenarios (Schwartz et al. 2023). Each stage poses its own challenges for students. For instance, the transfer from real-world phenomena or questions to the experimental space and then back to the real-world context when working with the results is tricky (Mönter and Otto 2016).

In addition to these three inquiry practices, modeling is often added to the list. Modeling involves simplified reconstructions of geographic reality. These models articulate central aspects and can be used to formulate explanations and/or make predictions .

Scientific inquiry practices entail physical and cognitive activities aimed at gaining understanding of the natural world (Schwartz et al. 2023). They are misunderstood if any activity involving movement, manipulation of materials regardless of cognitive engagement or target learning goals, is labeled as inquiry (Schwartz et al. 2023). Scientific inquiry practices encompass not only a hands-on approach, the skills and methods involved in scientific inquiry, but also a minds-on approach, the abilities to analyze and evaluate data in various representations, negotiate and argue from scientific evidence, draw appropriate scientific conclusions, critical thinking and decision-making (Xuan et al. 2019). Recent studies have shown that the time students spend conducting experiments is not necessarily decisive for their cognitive achievement. The minds-on phases before and after conducting an experiment are crucial for developing higher-order thinking skills and impacting cognitive achievement (Tesch and Duit 2004). Some researchers have extended the hands-on, minds-on analogy further to include a hearts-on approach. The hearts-on approach emphasizes emotional engagement with aspects such as enjoying the joy of science, feeling the beauty of science, experiencing the usefulness of science, and participating in the development of science. Context becomes more important for students to understand the societal relevance of science (Song and Cho 2004).

Students' inquiry skills need to be developed gradually (Tomcikova 2020). Depending on the learning group, the teacher should provide a suitable degree of instructional guidance, ranging from more structured instruction to guide students through the inquiry process to more independent and open approaches to support explorative learning processes (Matlen and Klahr 2013). Banchi and Bell (2008) offer an established trajectory concerning the degrees of students' freedom and teachers' guidance: confirmation, structured, guided, and open inquiry.

The third part of scientific literacy (3) includes an epistemological view, which refers to the understanding of what knowledge is and where knowledge comes from. These questions are addressed by the nature of science (NOS). This includes aspects such as learning scientific principles (e.g., the control approach) or engaging with the specifics of knowledge formation (e.g., hypothesize and test). NOS has been defined in numerous ways, but it most commonly refers to the values and assumptions inherent in scientific knowledge. This involves acknowledging that scientific knowledge (A) is tentative (subject to change), (B) is empirically based (derived from observation of the natural world), (C) is theory-laden, (D) necessarily involves human interference, imagination, and creativity (involves the invention of explanations), (E) necessarily involves a combination of observations and inferences, (F) is socially and culturally embedded, and finally (G) involves a combination of the function of and relationship between scientific theory and laws (Lederman 1992). Acquiring this perspective on science requires metareflection.

Approaches to Support Scientific Literacy in Geography Education

To address structural obstacles and enhance the scientific literacy of both teachers and students in geography education, two approaches are presented and discussed in relation to the challenges faced in science education within geography classrooms and scientific literacy. Both approaches are ready for dissemination: Either they can be directly used or replicated for other contexts.

Introduction to the Geobox Approach

The Geobox approach offers inquiry-based learning opportunities in a ready-to-teach format. It comprises materials and resources packaged in portable boxes along with a guidebook for teachers. During the development phase, the Geoboxes underwent iterative cycles of improvement (Fig. 1).

By analyzing the German Standards (German Geographical Society 2014) and other relevant educational documents, two overarching topics were selected: weather and soil erosion. These topics are highly relevant given the current global challenges and are frequently taught in schools. Moreover, they require a solid understanding of

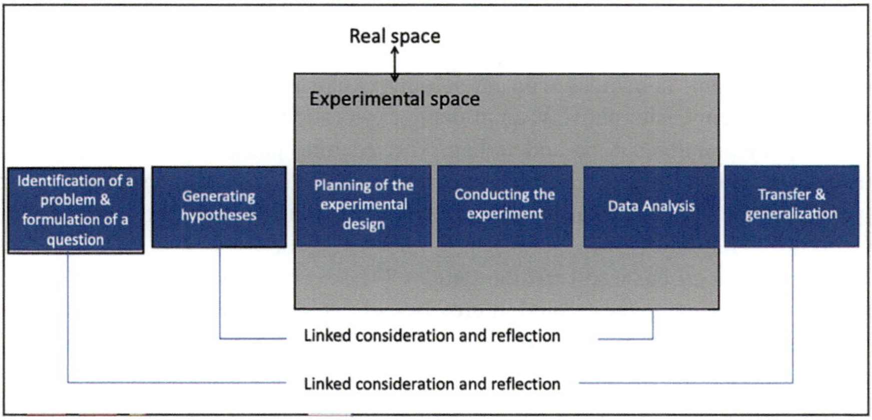

Fig. 1 Stages of an experiment (adapted from Rosendahl 2023 based on Otto and Mönter 2015)

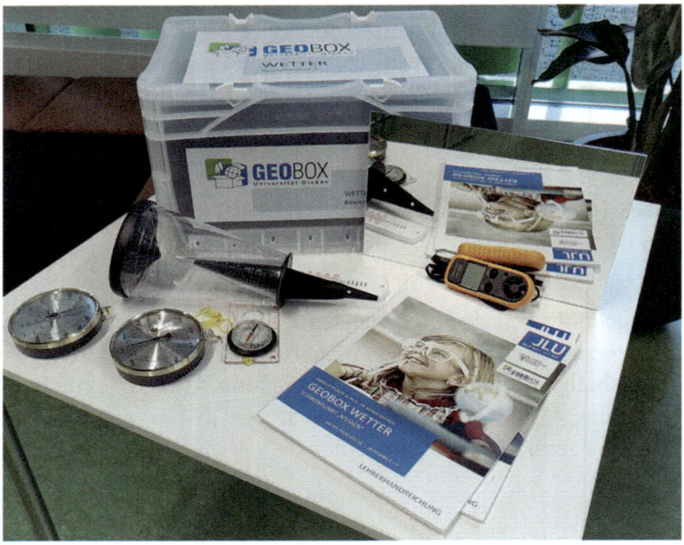

Fig. 2 Geobox 1 on weather elements

natural science principles. These topics were then integrated with practices of scientific inquiry, resulting in two Geoboxes: Geobox 1 "Measuring—Weather" (grades 5–7) introduces students to the measurement of various weather elements such as temperature, wind, precipitation, air pressure, and humidity. The Geobox provides the necessary devices for measuring these elements, including a barometer, thermometer, anemometer, rain gauge, and hygrometer (Fig. 2).

The teaching guide recommends starting with students' descriptions of the weather. Based on their descriptions, various weather elements are identified. Following this introduction, the teaching and learning process is organized into station-based learning. Students rotate through different stations, each focusing on a specific weather element, its corresponding measuring device, and the measurement process. The unit concludes with an activity exploring the interaction of various weather elements. Alternatively, an in-depth exploration of measurement as a scientific inquiry practice can be undertaken. The teaching guide includes an introduction to scientific literacy and measurement as scientific inquiry practice, lesson plans, student worksheets, answer keys, and instructions for replicating this specific Geobox.

Geobox 2, focuses on soil erosion (grades 8–10), centers on the experiment as the primary scientific inquiry method. It provides a learning opportunity to test different hypotheses related to soil erosion caused by water. The box includes various materials such as a watering can, a beaker, and wooden elements (see Fig. 3).

The teaching guide follows the structure of the different stages of an experiment (Fig. 1). It begins by posing the question of why water erosion affects one field while leaving another unaffected. A research question is formulated, and students are

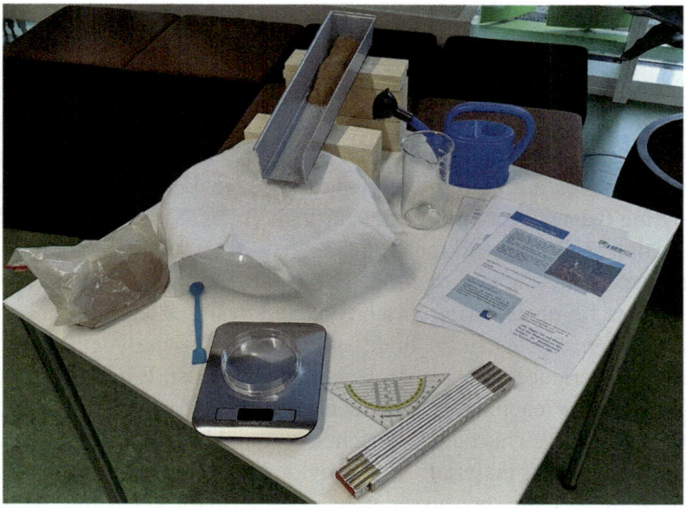

Fig. 3 Geobox 2 on soil erosion

tasked with generating different hypotheses. Next, students engage with the materials provided in the Geobox, select one of their hypotheses, and plan the experimental design. They conduct the experiment and analyze the results. The teaching guide includes all necessary worksheets and potential solutions as well as suggestions and alternatives to accommodate various teaching needs.

Additionally, it introduces the teacher to scientific literacy and the scientific inquiry method of experimentation. Furthermore, it provides students' results, including filled-in worksheets and transcriptions of their discussions while working with the Geobox, which can be analyzed and assessed in relation to students' preconceptions and common mistakes. Similar to the first Geobox, it offers detailed contents of the box for replicating it.

With the assistance of third-party funding, the Geoboxes approach also incorporates an innovative distribution and lending system. The boxes are stored in teacher training seminars distributed throughout Germany. These decentralized and widely available seminars serve as central hubs for borrowing Geoboxes. Four boxes of one type are stored in the seminar, allowing groups of four to five students to work on one box. Additionally, the seminars contribute to the sustainable establishment of the Geoboxes, as new generations of geography teachers are consistently trained there.

Due to the pandemic, four additional Geoboxes have been developed as digital offerings (visit www.geobox.online). These digital resources adhere to different principles due to the specific circumstances of the pandemic.

In summary, Geoboxes offer compact, comprehensive, and ready-to-use teaching resources, accompanied by support for teachers to facilitate science teaching effectively.

Discussion of the Geobox Approach

The Geobox approach addresses the structural constraints that hinder science teaching in geography classes. By providing all necessary resources for lessons, Geoboxes reduce the time teachers spend gathering supplies and planning instruction. These resources are conveniently assembled, making them easily accessible and eliminating the need for specific scientific facilities. Moreover, Geoboxes cover curriculum content. While working with Geoboxes does not require additional time, it does involve a different use of time. The Geobox approach offers flexibility in terms of the time frame, allowing it to be adapted to single lessons once a week.

The comprehensive teaching guide offers a user-friendly entry into science teaching for inexperienced geography teachers, at no cost. It provides fundamental principles of science teaching in a condensed format, focusing on specific topics and scientific practices. This enables teachers to quickly orient themselves and gain initial access to teaching materials. The provided solutions to students' worksheets not only reassure teachers but also emphasize alternatives and highlight essential aspects.

Both Geoboxes support scientific literacy by helping students gain knowledge about specific contents necessary to understand geographical issues. For example, Geobox 1 introduces students to weather elements and their interrelationships, while Geobox 2 provides insight into factors that accelerate soil erosion. The particular strength of the Geobox approach lies in its active engagement with scientific inquiry practices. Geobox 1 facilitates investigations and measurements, while Geobox 2 supports active experimentation, guiding students through the stages of an experiment. The teaching guide offers a learning opportunity that incorporates hands-on activities, such as regularly measuring and documenting data about wind speed or conducting experiments on soil erosion. Additionally, the Geoboxes foster a minds-on approach, prompting students to speculate about causes, formulate hypotheses, suggest experimental setups, analyze data, and draw conclusions. They are encouraged to generalize their findings and link them to the research question. A hearts-on approach was also considered, as the scientific phenomenon of water erosion was contextualized in a specific case example. Instructional guidance provided by the Geoboxes can be used to demonstrate experiments and encourage open experimentation by groups of students. While the teaching guide suggests explicitly discussing the stages of the experiment to help students understand what an experiment entails, it also considers epistemic features/functions and NOS on the sidelines.

A similar experimental kit was developed at Friedrich-Alexander-University Erlangen-Nuremberg, undergoing empirical testing regarding students' motivation and achievement when working with the experimental kits. The intervention study, conducted in a one-group pre-test/post-test design, revealed higher levels of practical activity and greater cognitive activation among students using the experimental kit compared to those in regular geography lessons. These findings indicate three ecologically valid functions of experimental kits: increases in students' activation, gains in content knowledge, and gains in knowledge about experiments (Naujoks et al. 2021; Schubert 2021).

Introduction to the Videographed Experiments Approach (ViExGeo)

The second approach to support science teaching involves utilizing short videos that feature geographical experiments, freely accessible through the link (https://www.youtube.com/watch?v=3bAjaN1AN68), many accompanied by a teaching guide (Rosendahl et al. 2024; Wieczorek et al. 2024). The project aims to go beyond the conventional use of videos by producing content specifically designed for integration into geography teaching. These videos are not intended to be self-explanatory tutorials or interactive virtual or augmented reality experiences. Instead, they showcase phenomena observed during experiments, allowing teachers to incorporate them into their geography classes. The videos, along with accompanying resources, provide various options for designing learning opportunities, which teachers can adapt to meet the specific needs of their students.

The videos demonstrate geographical experiments structured according to the stages of an experiment (Fig. 1). Since the videos are intended for classroom use, they skip the formulation of a research question and the generation of hypotheses, starting instead at the stage where the experiment needs to be planned. The videos are divided into three main sections. Firstly, viewers are introduced to the different materials. The materials are displayed to the camera and then set aside. Secondly, the next section illustrates the gradual setup of the experiment. Finally, the third section depicts the conduction of the experiment (Fig. 4), which is the longest segment. At times, certain parts are fast-forwarded or highlighted if they are time-consuming or difficult to discern. Additionally, the videos are intentionally kept basic to allow for flexible use in various contexts. For example, they are without sound to facilitate their use in international classrooms. Bookmarks are inserted at each section of the experiment for easy navigation. The following videos are available: water retention capacity of soils (https://www.youtube.com/watch?v=hf4lPjspgf8), soil erosion by wind with different influencing facotrs (grain size: https://www.youtube.com/watch?v=NWBVW-laOKo; windbreak planting: https://www.youtube.com/watch?v=kummPbhpFFg; vegetation: https://www.youtube.com/watch?v=JiuEJ0leQ30; wind strength: https://www.youtube.com/watch?v=aATmEyA_K6k) and soil erosion by water (vegetation: https://www.youtube.com/watch?v=VxgyP2sRFFI; precipitation amount: https://www.youtube.com/watch?v=0QptpfsUtpM; slope inclination: https://www.youtube.com/watch?v=4Jf53si9Cik). The platform is intended to grow progressively.

Each video is accompanied by an extensive teaching guide (Rosendahl et al. 2024). The teaching guide presents three different approaches to embedding the videos in instruction and facilitating teaching and learning science. Additionally, it includes

Fig. 4 Section 3 https://www.uni-muenster.de/Geographiedidaktik/schule/Unterrichtsmaterial/GeoBox.html

resources useful for all three approaches: an experimental log, sample solutions, and differentiation materials to accommodate different student needs. The most straightforward approach involves formulating the research question in a plenary session driven by some kind of impulse. Students' assumptions are gathered to answer the question, and selected assumptions are transformed into hypotheses recorded in the experimental log. Subsequently, students watch the entire video and record their observations in the experimental log. They have the opportunity to record and analyze data and measurement series with the help of the video. In the plenary, results are compared and discussed, and the teacher facilitates the transfer or generalization of the results/interpretation. The additional approaches offer methodological variants resulting from skipping parts of the video or stopping the video after a section. For example, one approach starts with the experimental setup (Sect. "(Insufficient) Science in Geography Education") and inquires which question could be answered with this experimental setup. Besides the three variants, multiple uses of the videos are possible, including in a blended learning environment. In the final section, the teaching guide offers content knowledge on the specific experiment, experiments in general, and a glossary with the most important terms in this context.

Discussion of the ViExGeo Approach

Considering the factors contributing to the neglect of science teaching in geography classrooms, the ViExGeo project offers a highly accessible entry point for both imparting and acquiring scientific literacy. It addresses structural constraints by requiring no additional equipment to integrate the videos into teaching. Moreover, both the videos and the accompanying teaching guide are freely available online without any special access requirements. These resources enable the presentation of lengthy experiments within a limited time frame, making the project highly efficient in terms of time and resources.

The teaching guide, in particular, is invaluable for geography teachers who may feel less confident teaching science. As mentioned earlier, it provides both content knowledge and pedagogical content knowledge specific to each experiment. The guide is well-structured and clear, offering a format familiar to teachers through ready-to-use lesson plans. However, as mentioned earlier, teacher agency is necessary to design suitable learning opportunities.

Regarding scientific literacy, it's worth discussing the extent to which the videos support scientific inquiry practices. The videos primarily focus on experimentation, which is considered the most complex and popular form of scientific inquiry. However, this approach neglects other forms such as investigation or modeling. Additionally, while real experimentation in the classroom is desirable, videos cannot fully replicate this experience. One major drawback is the absence of hands-on experimentation, which is crucial for students to develop skills such as pipetting. Furthermore, the popularity of experiments often stems from their deviation from so called "usual" (textbook) lessons. Using videos instead of hands-on materials may dampen students' motivation. Nevertheless, given the context of science-inexperienced teachers, it may not be realistic to expect them to conduct experiments with their students at all. In such cases, alternatives should be considered seriously.

Furthermore, Meier et al. (2022) have produced similar videos showcasing experiments for biology education. In this context, the videos are intended to supplement real experimentation. The researchers emphasize the usefulness of these videos for inclusive biology teaching. Learners can combine real experiments with excerpts from the videos, and the videos can be enhanced with audio tracks or text overlays. In an empirical study, biology teachers viewed the videographed experiments as meaningful and useful for their teaching, citing reasons such as time savings, availability of resources, danger mitigation, and the representation of complex experiments (Meier et al. 2022).

Conclusion

Integrating social and natural sciences is crucial for preparing society for the future and developing future-ready geography. It will equip future leaders with the knowledge needed to make well-informed decisions, understanding both science and human impacts and responses. Geography plays a unique role in educating upcoming generations to comprehend and balance these concepts (Selway 2021).

Geography education must contribute to scientific literacy, recognizing its significant potential in this regard (Xuan et al. 2019, p. 185). However, simply having this potential is insufficient; it must be effectively utilized. After analyzing the causes and exploring scientific literacy, this article proposes two pathways forward: Geoboxes, which focus on scientific inquiry practices, and ViExGeo, which showcases geographical experiments suitable for integration into geography teaching. The accompanying teaching guides are particularly beneficial for inexperienced geography teachers, facilitating their exploration of science education. These approaches offer numerous advantages for both teachers and students and can serve as valuable starting points.

However, we acknowledge that integrating a stronger emphasis on science into geography teaching requires more than providing kits, videos, and teaching guides. Effective science teaching necessitates teachers with deep content knowledge, pedagogical content knowledge, high confidence, and a strong identification as science educators (Jones et al. 2012). Achieving these aspects will bring about lasting changes in classroom teaching. While this is desirable, it is also complex and demanding. In the meantime, geography teachers can begin by exploring the approaches discussed above.

References

Akuma FV, Callaghan R (2019) Characterising extrinsic challenges linked to the design and implementation of inquiry-based practical work. Res Sci Educ 49:1677–1706. https://doi.org/10.1007/s11165-017-9671-x

Banchi H, Bell R (2008) The many levels of inquiry. Sci Child 46(2):26. https://doi.org/10.2505/3/sc08_046_02

Caldis S, Kleeman G (2021) Recognising geography as a partial stem. Subject: the journey so far. Geogr Educ 34:6–13

Eidietis L, Rutherford S (2009) Sailing toward understanding surface currents: a science and geography integration activity for upper-elementary students. Sci Activities: Classroom Projects Curri Ideas 46(3):5–14. https://doi.org/10.3200/SATS.46.3.5-14

German Geographical Society DGfG (Ed.). (2014) Educational standards in geography for the intermediate school certificate with sample assignments (3rd ed). Selbstverlag Deutsche Gesellschaft für Geographie

Grubbs ME, Grubbs S (2015) Beyond science and math: Integrating geography education. Technol Eng Teacher 74(4):17–21

Grube, C. (2010). Kompetenzen naturwissenschaftlicher Untersuchung der Struktur und Entwicklung des wissenschaftlichen Denkens bei Schülerinnen und Schülern der Sekundarstufe I. Universität Kassel

Hemmer I, Hemmer M (2010). Schülerinteressen an Themen, Regionen und Arbeitsweisen des Geographieunterrichts. Weingarten, HGD

Hof S, Hennemann S (2013) Geographielehrerinnen und -lehrer im Spannungsfeld zwischen erworbenen und geforderten Kompetenzen. Eine empirische Studie zur zweiphasigen Lehramtsausbildung. Geographie und ihre Didaktik 41(2):57–80. https://doi.org/10.18452/24615

Holbrook J, Rannikmae M (2009) The meaning of scientific literacy. Int J Environ Sci Educ 4(3):275–288

Höhnle S, Schubert JC (2016) Hindernisse für den Einsatz naturwissenschaftlicher Arbeitsweisen im Geographieunterricht aus Studierendenperspektive. Ausgewählte Ergebnisse einer empirischen Studie mit Lehramtsstudierenden. GW-Unterricht 142/143(2–3):153–161. https://doi.org/10.1553/gw-unterricht142/143s153

Jones G, Robertson L, Gardner GE, Dotger S, Blanchard MR (2012) Differential use of elementary science kits. Int J Sci Educ 34(15):2371–2391. https://doi.org/10.1080/09500693.2011.602755

Kruit P, Oostdam R, Van den Berg E, Schuitema J (2018) Effects of explicit instruction on the acquisition of students' science inquiry skills in grades 5 and 6 of primary education. Int J Sci Educ 40(4):421–441. https://doi.org/10.1080/09500693.2018.1428777

Lederman NG (1992) Students' and teachers' conceptions of the nature of science: a review of the research. J Res Sci Teach 29(4):331–359. https://doi.org/10.1002/tea.3660290404

Lethmate J (2005) Geomethoden. Kritische Anmerkungen zum fachdidaktischen Verständnis geographischer Arbeitsweisen. Geoöko 26(3/4):251–282

Lethmate J (2006) Experimentelle Lehrformen und scientific literacy. Praxis Geographie 36(11):4–11

Matlen BJ, Klahr D (2013) Sequential effects of high and low instructional guidance on children's acquisition of experimentation skills: Is it all in the timing? Instr Sci 41:621–634. https://doi.org/10.1007/s11251-012-9248-z

Meier M, Kastaun M, Stinken-Rösner L (2022) Experimentiervideos im naturwissenschaftlichen Unterricht – Lehren und Lernen mit und durch VidEX. In: Watts EM, Hoffmann C (eds) Digitale NAWIgation von Inklusion. Edition Fachdidaktiken. Springer VS, Wiesbaden. https://doi.org/10.1007/978-3-658-37198-2_5

Mönter L, Otto K-H (2016) Experimentelles Arbeiten im Geographieunterricht. Geographie Aktuell und Schule 219(38):4–13

Naujoks N, Gölitz D, Tellesch-Bülow C, Händel M, Schubert JC (2021) Students' motivation during experimental activities: an empirical study with GeoBoxes in Germany. https://doi.org/10.35542/osf.io/h6yvm

Oldakowski R, Johnson A (2018) Combining geography, math and science to teach climate change and sea level rise. J Geogr 117(1):17–28. https://doi.org/10.1080/00221341.2017.1336249

Orion N, Libarkin J (2023) Earth science education. In: Lederman N, Zeidler D, Lederman J (eds) Handbook of science education. Routledge, pp 692–716. https://doi.org/10.4324/9780367855758

Otto K-H (2003) Experimentieren Im Geographieunterricht. Geographie Heute 24(208):2–7

Otto K-H, Mönter L (2015) Scientific literacy im Geographieunterricht fördern Experimentelle Lehr-/lernformen und Modellexperimente. Geographie Heute 36(322):2–7

Peter C, Sprenger S (2014) Kompetenzgrad von Lernenden beim Experimentieren im Geographieunterricht. Zeitschrift für Geographiedidaktik 42(3):179–191. https://doi.org/10.18452/23988

Rosendahl N, Wieczorek L, Rüther L, Scholten N, Bienert N, Mehren R (2024) Geobox: Digitales experiment—Bodenerosion durch Wasser. Lehrkräftehandreichung. https://www.uni-muenster.de/Geographiedidaktik/schule/Unterrichtsmaterial/GeoBox.html

Rosendahl, N. 2023. Geographiedidaktische Forschungen, Bd. 80, Experimentieren im GEO Lehr-Lern-Labor. Eine DBR-Studie zur Förderung der Selbstwirksamkeitserwartung von Studierenden zum naturwissenschaftlichen Arbeiten im Geographieunterricht N/A: Selbstverlag / Eigenverlag.

Schneegans S, Nair-Bedouelle S (2021) Scientific literacy: an imperative for a complex world. UNESCO Publishing, Paris

Schubert JC (2021) Developing geographic skills through experiments: implementing experiments in geography classroom through geoboxes. Rev Int Geogr Educ Online 11(2):525–539. https://doi.org/10.33403rigeo.851222

Schwartz R, Ledermann J, Enderle P (2023) Scientific inquiry literacy. The missing link on the continuum from science literacy to scientific literacy. In: Lederman N, Zeidler D, Lederman J (eds) Handbook of research on science education. Routledge, pp 749–782. https://doi.org/10.4324/9780367855758

Schwippert K, Kasper D, Köller O, McElvany N, Selter C, Steffensky M, Wendt H (2020) TIMSS 2019: Mathematische und naturwissenschaftliche Kompetenzen von Grundschulkindern in Deutschland im internationalen Vergleich. Waxmann Verlag. https://doi.org/10.25656/01:21353

Selway K (2021) STEM in geography education—an earth science perspective. Geogr Educ 34:18–21
Song J, Cho S-K (2004) Yet another paradigm shift? From minds-on to hearts-on. J Kor Assoc Res Sci Educ 24(1):129–145. https://doi.org/10.1142/9789812702890_0004
Tesch M, Duit R (2004) Experimentieren im Physikunterricht—Ergebnisse einer Videostudie. Zeitschrift Für Didaktik der Naturwissenschaften 10:51–69
Tomcikova I (2020) Implementation of inquiry-based education in geography teaching—findings about teachers' attitudes. RIGEO 10(4):533–548. https://doi.org/10.33403rigeo.791713
Vasileva M, Poleganova D (2017) Conducting experiments in geography training twelfth international scientific conference: knowledge without borders, Vrnjacka Banja, Serbia
Wieczorek L, Rosendahl N, Mehren R (2024) Experimentiervideos: Bodenerosion druch Wind Lehrkräftehandreichung. https://www.unimuenster.de/imperia/md/content/geographiedidaktik2/materialfuerschulen/geobox_winderosion_lehrkraftehandreichung_wieczorek_rosendahl_mehren-1.pdf
Xuan X, Jin Q, Jo I, Duan Y, Kim M (2019) The potential contribution of geography curriculum to scientific literacy. J Geogr 118(5):185–196. https://doi.org/10.1080/00221341.2019.1611906

Discussion Groups in Social Media as a Tool for Detecting Critical Educational Events: The Example of Polish Teachers of Geography

Kinga Bargieł

Abstract In recent years, schools have been working in conditions of permanent uncertainty resulting from numerous critical events that have occurred in the field of education. Teachers have always consulted each other as to possible ways of meeting the challenges they face resulting from systemic, political, and social changes. The widespread access to the media resulted in social media becoming an area for this type of discussion and seeking support. Today, closed online discussion groups are a place for teachers to exchange experiences, opinions, and solve problems. It can be noticed that groups associating teachers (both general and subject groups) have been operating particularly dynamically for a long time and are constantly developing. Closed discussion groups are able to gather tens of thousands of teachers from all over Poland who publish hundreds of educational posts a day. Recent critical events that pose a challenge to most teachers have generated problems and reasons for uncertainty. The aim of the proposed chapter is to analyze the content of entries in the "Teachers of Geography" social group (Facebook) in order to recognize the key problems that teachers of Geography have faced and continue to face nowadays as a result of critical events. The chapter will carry out both a quantitative and qualitative analysis of the entries. The quantitative analysis has showed that at moments of critical events such as remote or hybrid learning resulting from COVID-19 pandemic, the war in Ukraine and the widespread appearance of foreign-language children at schools, new ministerial regulations, the number of posts and queries published was much higher than before. In turn, an examination of the content of the entries has showed that the topics of the published posts have been varied and they have mainly involved requests for inspiring teaching materials covering almost every area of teachers' teaching work (both authored and those found on the Internet), provisions of educational law related to current school challenges, as well as conferences and trainings that have supported them in meeting subsequent teaching and educational challenges.

Keywords Social media · Critical education events · Teachers of Geography · Discussion group · Facebook

K. Bargieł (✉)
University of the National Education Commission, Podchorążych 2, 30-084 Krakow, Poland

© The Author(s), under exclusive license to Springer Nature Switzerland AG 2025

D. Piróg and A. Świętek (eds.), *Future-Ready Geography*, Key Challenges in Geography, https://doi.org/10.1007/978-3-031-81268-2_6

Introduction

The basis of contemporary interpersonal communication proves to be contacts through social media (Fabjaniak-Czerniak 2012). This way of conducting virtual social relations has become particularly widespread on account of an easy access to Internet. Additionally, recent events (pandemic, war) have intensified the frequency of using this type of media for various purposes. These include not only communication between people but also their seeking answers to questions from various areas of their interests, both personal and professional (Majewska 2021). There has been an observable trend where users primarily attempt to find the answers to their questions in the sources in virtual reality rather than, for example, using specialized literature, thematic publications, or direct conversations with people. While undertaking these actions, they first and foremost tend to utilize the resources of the Internet, searching for the latest answers to their questions (Wójtowicz 2019; Iwanowska 2016). In this context, discussion groups have become particularly fashionable, where members of the groups, for example, those residing in a particular neighborhood, studying the same specialty, or belonging to the same professional group, join together. These forums can be multi-themed or related to specific, detailed, or even specialized topics. Their formula can be both open and closed. As for open groups, every user has their own access to active discussions, whereas the access to closed groups has to be obtained only through a designated verification key. As for the discussion groups, users can add posts, ask questions, provide answers, comment on posted statements, or even react non-verbally (Pietrzak-Zawadka and Lewoń 2018). The aim of discussion groups is the virtual integration of interested users, offering mutual inspiration, or exchanging their own experiences. Research indicates that social networking sites and social media are becoming increasingly popular means of communication for representatives of various professional and social groups (Janc and Jurkowski 2020; Perz 2022). Taking into account the characteristics of online discussion forums, their content can function as a valuable source of information about the current situation, problems, and opinions of the members gathered therein. Therefore, they have become appreciated sources of data for scientific research within disciplines such as psychology, sociology, or marketing (Łosiewicz 2013).

Due to the fact that social media is a relatively new phenomenon, its research potential is currently both being discovered and increasingly developed. Research opportunities of social media discussion groups have also been recognized as valuable sources of information as far as various disciplines are concerned. Therefore, research procedures are being developed for the scientific analysis of the content posted in discussion groups.

The identified characteristics of discussion groups can provide relevant information necessary for identifying critical events in education, namely those that occur unexpectedly and suddenly and have a significant impact on the educational process. There are analyses of social media discussion groups concerning various aspects of education, for example, those related to early childhood teachers or flipped classrooms. However, research material indicates that they have not so far focused on forums for Geography teachers.

In this chapter, both the procedure for content analysis of discussion forums and the results of the content analysis regarding the recognition of critical event records in terms of their frequency and temporal dynamics of entries were presented. The subject of the research was the Polish discussion group "Geography Teachers," which was established in 2017 and has 4300 members (as of March 25, 2024). According to studies of the subject, no research has been conducted so far that involves analyzing the content of discussion groups bringing together Geography teachers in any context. Additionally, no research has been conducted aiming to identify critical incidents based on the posts made in such groups. Meanwhile, the content posted by its users is extremely valuable for research purposes because, unlike any other source, it continuously reveals the current problems and educational challenges that Geography teachers are experiencing and dealing with. Analyzing these posts not only allows for the diagnosis of the current situation at schools but also for forecasting problems that may arise in the future. Monitoring the topics of the posts, including the reactions of users after their publication, can become a helpful guide (for researchers, the School Board, the Ministry, among others) in diagnosing current educational problems and challenges and addressing them in their early stages.

Geography is one of the school subjects where the content topics dynamically undergo changes. The economic or geopolitical situation of the world, climate change, demographic indicators, and average monthly rainfall are just examples of issues that are constantly evolving. The dynamic changes in socio-economic and physical Geography mean that topics discussed by Geography teachers during their lessons quickly become outdated. Therefore, Geography teachers ought to/must continually revise and verify their knowledge. The rapid advancement of modern technologies, including GIS environments, is closely linked to modern geographic education. Monitoring and diagnosing the needs related to the workshop and knowledge sources of Geography teachers will help improve the level and quality of geographical knowledge they impart.

Like all Internet users, Geography teachers seek the latest and most reliable sources of information on online forums. They look for specialized content in forums and groups that gather teachers of particular subjects. The diversity of school subjects means that teachers face different problems and challenges in their respective fields. Geography teachers, for instance, will not be searching for source texts ideal for interpreting the linguistic styles of a particular era. The topics of their posts will revolve around field trips, maps, diagrams, and current demographic data.

The analyses focused on diagnosing critical events that occurred between 2020 and 2023 and resulted in intensified discussions among group members on various topics related to their profession—Geography teachers.

A critical event is an interdisciplinary concept, variously defined. The common denominator understood by researchers representing multiple scientific disciplines is that it refers to an event or incident that occurred unexpectedly and suddenly. The assessment of a particular event as critical always occurs subjectively, meaning individually by a given person. In the literature on the subject, a synonym for this term is also used—critical incident (Zarzycka 2010).

This concept is particularly widely analyzed and interpreted in psychology and social pedagogy; on the one hand, as events that have changed an individual's life, and, on the other hand, as a process that has influenced taking actions related to eliminating difficult life memories. It is also called a "turning point in a person's life." It is an event that "appears suddenly, with great force, causing a disruption of the individual's previous stability in one's life, thereby motivating them to take particular actions" (Krawczyk-Bocian 2013, p. 26). Critical event is understood as a critical life event because they fully relate to the emotional sphere of a human being. They can play a role as both a positive or negative event that can lead to the development or regression of the individual (Kuryś-Szyncel and Błachnio 2020).

Due to the fact that the study, the results of which are presented in this chapter, encompasses the educational aspect (in relation to Geography taught at schools), in this chapter, the author adopts a definition with a pedagogical-sociological character offered by David Tripp and Sabina Cisek. These authors understand a critical event as:

> a single event (or situation) that can be distinguished and significantly influences some action or phenomenon (Cisek 2017: p. 8)

and

> an event or situation that becomes a significant turning point in the fate of a person, an institution, for example, a political party, or during some social process such as industrialization, war, negotiations. (Tripp 1996, p. 16)

In both definitions, the authors focus on the effects of critical events, which are important and significant for forecasting changes and consequences related to the functioning of society. The individual categorization of an event as critical is associated with its impact on a particular individual/institution or professional group. An example could be changes in the school environment. For retirees or people not associated with education, information about the introduction of remote learning may not be perceived as a "critical event." However, for the teaching or student community, it certainly can be.

Discussion Groups on Social Media

The primary source of the latest pieces of information from the world of politics, sports, and education has become social media. They are gradually taking over the role of news portals or television. For some people, they serve as diaries or "Virtual Hyde Parks"—places designed for expressing their own opinions and conducting discussions. The intensity of published data and user activity related to a particular topic can indicate its growing importance to people. Members of social virtual groups react to topics that affect their lives or fall within their interests (Naruszewicz-Duchlińska 2005). While focusing their attention upon thematic groups, also known as discussion groups, users actively exchange their opinions on topics that interest them. Groups function as the "cafés" of social media because they bring together smaller groups of concerned individuals who can interact with one other, activating themselves and writing about issues important to them (Lupa-Wójcik 2020). They create a virtual space for exchanging views and conducting correspondence on a given topic (Juza 2020).

It might seem that online discussion groups emerged thanks to social media messengers in the early twenty-first century. However, it turns out that these groups have had much longer history. The USENET system was established as early as 1979, being dedicated to discussions among users worldwide (Kućko 2019). In the 1980s and 1990s, USENET was the most popular application for conducting group conversations. Discussion groups were divided hierarchically—from general topics to more specific content. On the Internet, there were other, less popular platforms for group discussions, but they usually operated within closed structures within one university. At the beginning of the twenty-first century, with the development of social media, USENET began to lose its popularity. Its position was largely taken over by social media, one of the main functions of which is discussion groups. For Generation Z, the largest population among today's Internet users, USENET seems to be completely unknown. They were born at the peak of social media platforms. Nevertheless, discussion groups are very popular among users of the "young generation" (Paczka 2020). We can access discussion groups on social media platforms such as Facebook, X (formerly Twitter), Reddit, or Instagram (as a broadcasting channel).

Social networking sites such as Facebook, Instagram, TikTok, and X are well-known to everyone. Social media play a function of Internet portals whose main purpose is to facilitate interactions between users (Grębosz et al. 2016). Among these, the most popular social networking site is Facebook, which in January 2024 was actively used by over 3 billion people worldwide (2023 META Report) (Fig. 1).

The popularity of Facebook can be confirmed by the slogan "If you are not on Facebook, you do not exist." The platform was created as a means of communication between individual groups of people from one university (Harvard). No one expected that it would become the beginning of a new era in global communication (Forst 2017; Wójtowicz 2019). In addition to posting, commenting, and texting, one of the most important features of Facebook is its discussion groups. There are no published data regarding the number of groups, but their popularity seems to be immense. There are groups whose members discuss all possible topics. Users create smaller groups or circles around their interests, bringing together local communities or professional groups. The undeniable fact is that the Internet has given us the opportunity to search for information on any topic within seconds, but in order to find the latest information, users tend to rely on discussion groups. In these groups, members can add content, ask questions, or post photographs, while other members can comment, express their opinions, or share their own experiences (Bąk 2016). In recent years, there has been a significant increase in activity among users of discussion groups and social media

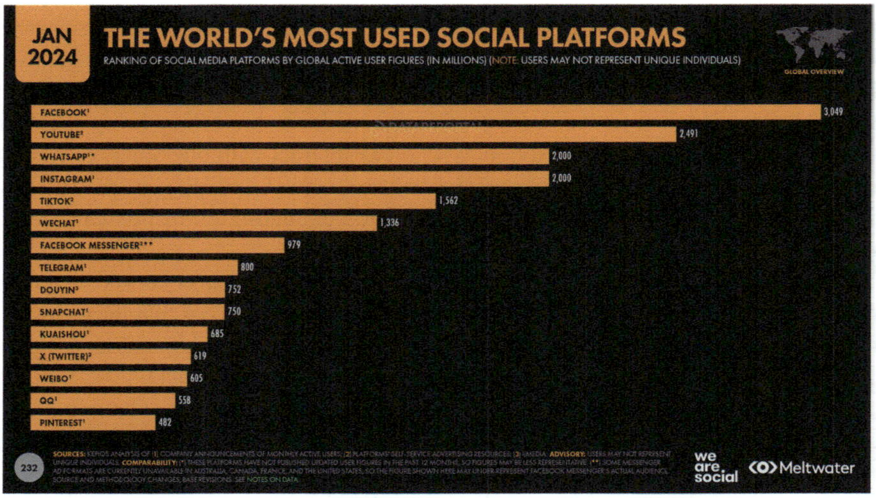

Fig. 1 World's most used social platforms in 2024 (*Source* Data Reportal 2024)

(Jaskiernia 2021). Social media has become particularly important for teachers and school communities. Due to unexpected critical events that occurred in Poland in the last three school years (e.g., remote or hybrid learning, the war in Ukraine, and the widespread presence of foreign-language children at schools), teachers have been forced to change their teaching methods and resources. Social networking sites and social media have become very popular means of communication among students, parents, and teachers alike (Chęcińska-Kopiec and Ochoa-Siguencia 2020). It can be observed that for some time now, groups uniting teachers (both general and subject-specific groups) have been particularly active and flourishing (Bergviken-Rensfeldt et al. 2018). Closed discussion groups bring together tens of thousands of teachers who publish a wealth of educational posts daily. The recent critical events, posing a considerable challenge for most teachers, have compounded problems and reasons for uncertainty. The ubiquity of access to media has led to an even more intensive turn by teachers toward social media, where they both share their professional dilemmas and seek help among other members of their profession (Al-Douri 2021).

Methods of Social Media Research: A Review of Studies of Subject

Since their inception, social media have been recognized as significant sources of social data, providing insights into users' daily lives (Wąsowicz-Zaborek 2020). Due to the diversity of data conveyed, they are of interest to almost every type of industry, particularly marketing, commerce (e-commerce), and advertising. They also serve as data source for researchers studying sociological or psychological processes.

User-generated content on social media is highly popular among researchers. Due to the various forms of content published (posts, photographs, videos), there are many ways to analyze and interpret them. Internet marketing specialists identify up to 100 types of data offered by social media that can be used for marketing analyses (Płuciennik 2014). The primary source of data includes all content published by users, as well as their activity related to other people's content (comments, likes, views, types of reactions, shares, or visits to the main profiles). Content undergoes multidimensional qualitative and quantitative analyses. Wolny (2017) defines several dimensions of social media analysis: the textual data analysis, the user analysis, the social network dimension, the geographical dimension, and the graphical dimension. Depending on the goals, needs, and interpretation, statistical models such as Text Mining, Latent Dirichlet Allocation (LDA), or topic and trend analysis are advised to be used.

An area of significant research interest lies in the individual behavior of Internet users. Issues related to people's activities online allow for monitoring social changes. Studying posts published online on public profiles and analyzing them in terms of expressed moods and emotions allows for investigating differences in behavior on social media in this regard (Thelwall et al. 2009). There are also studies related to the analysis of users' profiles, pieces of information contained therein, and their network of friends. Example studies included data regarding the type of account (private or open), relationship status, gender, religion, education, or nationality of the users. Data obtained allowed for, among other things, exploring the perception of privacy on social media and identifying differences among generations using these communication channels (Thelwall 2008; Levickaite et al. 2010).

Social media are also studied in terms of comparing the content published by institutions dealing with similar topics (e.g., university libraries, universities) (Przybysz and Pioterek 2015). Control studies have also been conducted, involving the publication of posts in various forms (a video, a post, a link, and a photograph) from the profiles of several institutions (e.g., hotels). The aim of the study was to determine which type of content generates positive attitudes among people and what the marketing effects of using social media really are. Survey research was used, which is the simplest form of research in mass media (sending surveys to users) (Leung 2012). There are studies in which social media profiles of various institutions from the same sector (e.g., universities from a particular region) were analyzed. Data concerning the gathered group, an account creation date, dynamics of fan growth, and the quantity, and type of content published were collected.

Research has shown that students demonstrate the greatest interest in social media. Therefore, the activity of university-related accounts associated with student life is particularly important (e.g., university social media accounts). Sample studies have revealed that private universities exert greater efforts in developing their social media profiles, as they publish more content and it comes in more diverse forms (Rynkiewicz 2011). Social media research conducted by linguists focuses on the characteristics of individual social media accounts or discussion groups in terms of specific types of literary forms or language used (Kulesza-Gulczyńska 2016). Ethnographers explore ethnic, national, or ethnographic minority communities through social media. Their ethnographic method involves observing posts (e.g., regarding language used online) and activities on social media. This method helps determine the level of attachment individuals have to their communities and the vitality of language among young users (but not only) (Dołowy-Rybińska 2017).

Depending on the social networking site, the process of acquiring data can be either easier or more difficult. Social media platforms that require login credentials for the access are less popular data sources. Despite having the largest number of users, Facebook is not commonly used for social analysis by the e-commerce sector. The access to individual social information is limited to logged-in users or those belonging to the friends' circle. Obtaining data from closed discussion groups, as the name suggests, proves to be particularly problematic. These groups are closed off to a wider audience. These difficulties result in discussion groups being the least researched area of interest (Wąsowicz-Zaborek 2020). However, the closed nature of these groups yields positives in terms of the authenticity of posted content and the members of the groups being selected through verification.

Researchers utilize thematic groups available on social media platforms to disseminate thematic surveys among their members. This facilitates reaching the target group of respondents (e.g., in the context of what they expect from groups—marketing research) (Lupa-Wójcik 2020).

Research related to discussion groups on social media is typically limited to a single thematic thread. An example could be research on the use of social media by teachers, indicating a growing awareness of significance of discussion groups for them. Studies focusing upon a group of Swedish teachers regarding the topic of "Flipped Classroom" concentrated their attention on the group's development over time, the number of active users, and survey research concerning reasons for belonging to the group (Bergviken-Rensfeldt et al. 2018). Similar survey research focused upon the Polish group of Art teachers or supporters of forest education (Al-Douri, 2021; Pietrzak-Zawadka and Lewoń 2018).

Research related to early childhood education teacher groups involved asking questions related to the benefits of belonging to the group and reflections on its functioning. Additionally, the author identified the themes of posts appearing in the group's resources (Marek 2017). A study conducted in Poland based on "Me, the Teacher" group involved analyzing and comparing the content of posts from two periods. The aim was to demonstrate changes regarding the themes of published posts (Kubica 2022). Research related to the school environment likewise addressed the use of social media during students' performing group tasks. The basis of the study was a survey directed at students concerning the use of media during their group work (content sharing, discussion groups, correspondence exchange, etc.) (Razmerita and Kirchner 2014). Social networking sites often serve as places for research and observation of students. Researchers create discussion groups (in which they are administrators) related to the subject matter of one particular subject or consisting of students from a particular year, specialty, or even faculty. This allows them to monitor processes and behaviors among users (Athanassopoulos et al. 2021). Nicholas Bowman and Mete Akcaoglu (2014) conducted research by creating a discussion group dedicated to one subject for students of a particular year. Both students and instructors published content related to the subject matter. The aim was a content analysis, but research also showed that students belonging to the subject group had higher final grades than classmates who did not belong to the group. Interesting research involves comparing Facebook discussion groups dedicated to a number of topics in books. The study involved gathering information about group access, the number of members, and analyzing the content of posts. Like in other studies, a questionnaire was used, focused upon users, containing questions about the reasons for belonging to the group and the benefits derived from it (Kosik 2017).

Discussion groups, especially those closed ones, are scientifically valuable and unique in terms of content, serving as sources for scientific analyses. The dynamically changing environment associated with Geography remains unexplored. In addition to recognizing the current state, the study of discussion groups can provide us with future perspectives. Both the descriptive and prospective aspects ultimately convince us to utilize forums for scientific analyses.

Research Procedure

To understand the current issues and challenges faced by Polish teachers of Geography and to identify critical events among them (see the subsection Critical event detection), the analysis was conducted on user posts published between March 1, 2020, and March 31, 2023, spanning 36 months. This constituted a complete sample, totaling 8482 entries, containing pieces of information such as the publication date, the number of comments and user reactions, as well as the full text of the post.

Given that the most important element of discussion groups is the verbal posts of users, research focused on the content of these posts, both longer and shorter expressions of its members. The content analysis allows for an objective, quantitative, and systematic description of the content (Klepka 2016). The analysis of teacher posts content was conducted using both qualitative and quantitative methods, employing a categorical key, a frequency analysis, and elements of holistic observation.

The first stage of the analysis involved observing the emerging posts, which allowed for preliminary conclusions regarding the topics of the posts and user activity. Subsequently, a complete sample of posts was collected, taking into account their content, publication dates, number of comments, and user reactions. The observation and gathering of the entire material enabled the creation of a classification key (Chromińska and Raciborski 2019), the creation of which was conducted in two stages. The observation assisted in developing a preliminary, a priori classification key based on existing knowledge. This helped in determining basic keywords and creating main categories. The main keywords (and their synonyms) used for classification into themes were as follows: *seek, share my content, recommend, need,* and *boast*. In the next stage, individual verification of the content of each post took place, allowing for the creation of a final classification key consistent with the emergent categorization technique. In total, 22 categories were identified. It was assumed that the categories in the key were mutually exclusive, so each post could be assigned to only one category, and the categorization was reliable (Szczepaniak, 2012) (Table 1).

After having compiled the database and the full classification of content, both a qualitative and a quantitative analysis became possible. The observation and a qualitative analysis of the published content helped determine the main themes of the emerging posts. While using a timeline, periods of an increased user activity were identified in terms of the number of posts added, which revealed reliance related to the occurrence of critical events during the study period. A quantitative analysis using frequency also allowed for determining the topics that were most frequently discussed and those that generated the most reactions or comments from users. The final step of the procedure involved the actual analysis, aiming to identify critical events and problematic areas based on the published content both qualitatively and quantitatively (Fig. 2).

Table 1 Classification key with categories

Number of category	Thematic scope
1	Sharing original teaching materials
2	Boasting children's projects and lesson outcomes
3	Searching for teaching materials and inspiration
4	Recommendations/expressing opinions on textbooks, atlases, or books
5	Work proposals for teachers
6	Teacher–student relationships
7	Recommendation of websites/articles/videos
8	Sharing original photographs and travel experiences
9	Advice related to remote teaching
10	Legal advice regarding teacher career advancement, teacher cards, regulations
11	Field trips/excursions—inspiration and suggestions for places and guides
12	Professional, methodological, and assessment-related advice
13	Proposals for innovations and projects
14	Equipping Geography classrooms
15	Training sessions, conferences, webinars
16	"Geography teachers after hours"—geographical jokes, memes, or creative student responses at exams
17	High school Matura exams/Matura exam tasks
18	Geography competitions and Olympiad
19	Joint actions and inter-school cooperation
20	Sharing PDF files of articles/books/work sheets
21	Recommended exercises from the internet
22	Buying/selling/giving away items

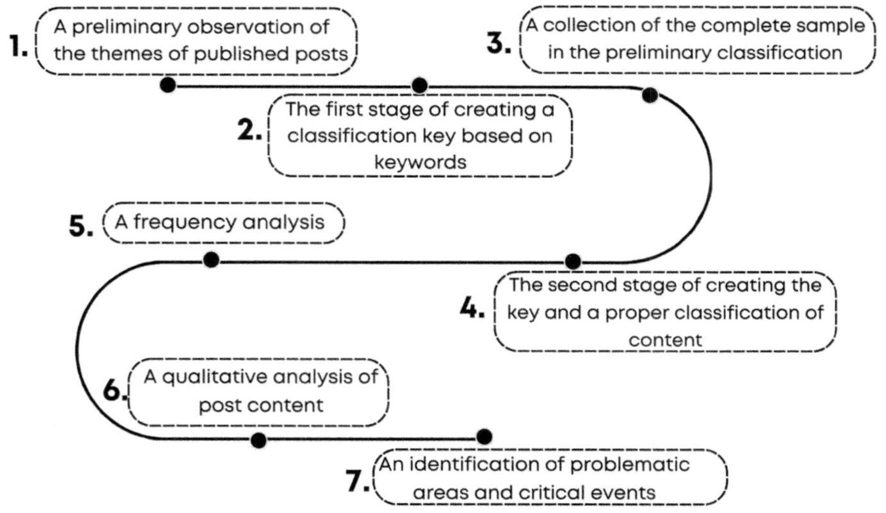

Fig. 2 Stages of research

The developed and applied research procedure takes into account the previous findings regarding the textual content analysis. Most of the known methods so far have mainly focused on posts in publicly accessible social media channels, which were not sufficient to achieve the research objective. Therefore, regarding the research subject, the research procedure is innovative, as previous studies did not concern Geography teachers and did not attempt to identify critical events in this way.

Results

The Content Analysis

The observation and application of the classification key allowed for the identification of the most frequently discussed issues in the published geographical posts. The content analysis of these posts led to the determination that they predominantly feature topics related to (Fig. 3):

- sharing of original teaching materials (presentations, exercises, lesson plans);
- sharing of materials/articles/websites/videos/photographs found online, related to Geography, which teachers can use in their lessons;
- inquiries directed to users asking for help/opinions/advice regarding, among others, curriculum implementation or formal school documentation;
- information about professional development training and educational conferences;

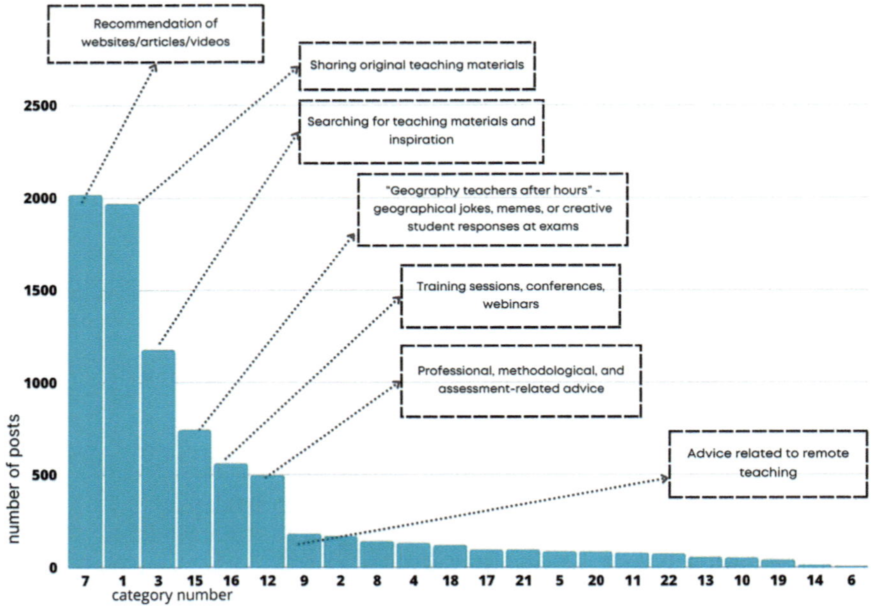

Fig. 3 Dominant themes of posts published in "Geography Teachers" group during the period of 01.03.2020–31.03.2023

- dilemmas related to assessment, methodological consultations regarding task-solving, reflections, or professional advice from experienced teachers;
- informal "teacher-related" topics, such as remarkable students' answers during exams, geographical purchases, or jokes.

During the analyzed period, a total of 8482 posts were published, averaging approximately 7.5 posts per day. The majority (over 2000 posts) was related to fascinating and practical materials such as diagrams, articles, photographs, or videos found on the Internet, specifically related to geographical education, which could enrich Geography lessons. These shared contents also included methodological materials regarding the teaching process and further training for teachers in the application of methods and teaching aids, as well as the use of modern technologies in Geography education. Slightly fewer posts (1970 entries) were noted regarding original teaching materials. Geography teachers shared their multimedia presentations, worksheets, tests, or ideas for engaging lesson activities. These posts were received with the greatest enthusiasm and gratitude, evidenced by the fact that they generated the most comments and reactions from the group's members. During the analyzed period, posts about original materials garnered over 42,000 user reactions (which accounts for over 38% of all published posts) and 9600 comments, constituting 20% of all comments posted by users from March 1, 2020, to March 31, 2023. These two thematic areas together represent approximately 50% of the posts that appeared during the study period.

When preparing lessons about soils, I was greatly disappointed when searching for appropriate profile diagrams on the Internet. Either they were crap or they didn't exist at all...
That's why, together with my main illustrator (student), we decided to solve this problem!

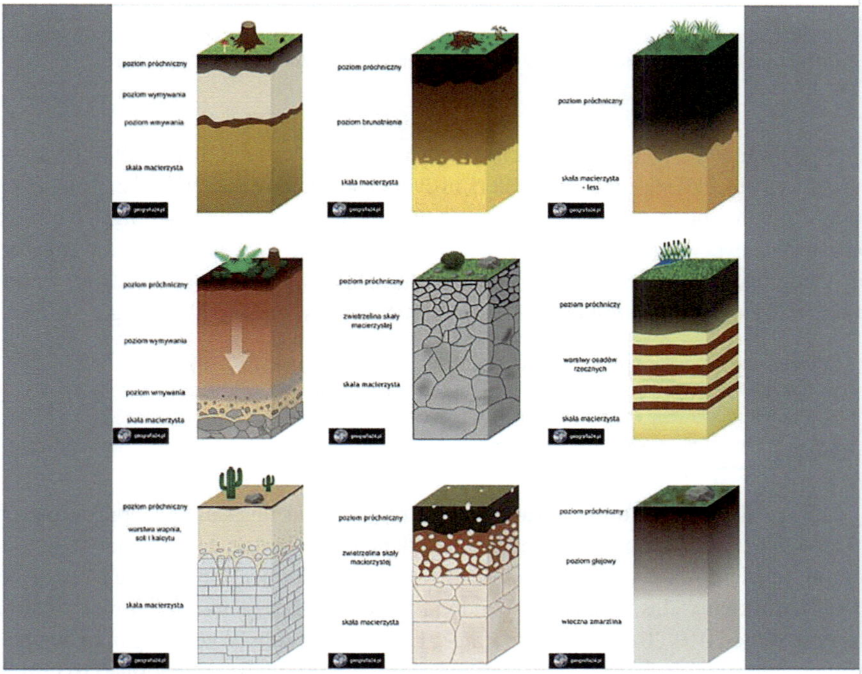

[redacted]i and 192 other users 30 komentarzy

The discussed discussion forum also serves as a place for seeking ideas and inspiration. As much as, 15% of the posts were requests to join forces and share experiences in order to collectively design motivating lessons or exercises. In posts of this character, there is a phenomenon of brainstorming where collaborative creative efforts often result in innovative and unprecedented materials.

I'm looking for inspiration for a lesson about savannahs in grade 5. Something that will interest, inspire, or possibly make the guests laugh (read open lesson).

4 9 comments

The Geography teachers' discussion group has also become a source of information regarding training sessions and conferences dedicated to Geography teachers, as well as events promoting Geography among students (webinars for students). The group gathers Geography teachers with diverse levels of professional experience. As it serves as a meeting place for "expert-practitioners" with teachers who are at the beginning of their career, a portion of the posts (6%) offers pieces of advice related to the profession or even reflections. Many of these posts are of a consultative character (e.g., regarding task solutions) or dilemmas related to assessment.

As a teacher with short experience, I teach in a primary school, I still have a lot of doubts about how to set priorities in my work. It seems to me that there is too much material, children do not understand it and are discouraged from learning geography, even though we know what a super interesting subject it is. And when implementing the core curriculum, at the end it turns out that the kids don't know the basics. They have problems with, for example, the directions of the world, the location of continents or oceans, not to mention countries. IN ... **View more**

5 35 comments

The content analysis of the dominant themes in the posts indicates that the group serves as a place of support for teachers in every aspect. However, it is also a place that brings together people with the same passion for Geography and teaching. Therefore, on average, every 14th published post can be thematically classified into the category of "Geography Teachers off-duty," which relates to the teaching practice from within, students' responses during exams, geographical jokes, or even shopping.

In the morning, a person opens the librus and there is this information from my student.
"I would just like to wish you all the best for the beginning of the new hydrological year! May the glaciers melt moderately and you will never run out of water in your tap."

[redacted] and 154 other users 13 comments

During the period under examination, remote learning lasting approximately 15 months was announced multiple times depending on the educational stage and region. In March 2020, by decision of the Ministry of Education and the Ministry of Health, stationary classes were suspended for the first time. This situation lasted until the end of the 2019/20 school year (June). The 2020/21 school year began in a hybrid manner, but schools from various regions of Poland gradually transitioned to remote learning. By November 2020, all schools across Poland were conducting classes remotely. This mode of learning persisted until the end of May 2021, when students returned to stationary education for a month until the end of the school year. Remote learning also occurred during the 2021/22 school year. From December 2021 to the end of February 2022, students of upper grades of primary and secondary schools continued learning remotely.

Guys, I have a question. Will you offer online lessons? Will you rely only on e-textbooks, sending regular and interactive tasks, presentations, and completing quizzes online? What does it look like for you?

1 41 comments

As a result, one of the more popular topics was related to conducting Geography lessons in a non-stationary format. Posts focused strictly on pieces of advice regarding remote teaching (e.g., conducting lessons, assessment principles, applications, and software). It was noticed that the period of remote learning was a critical moment in the careers of some teachers because the content of the posts indicated that they were facing difficulties with basic issues related to Information Technology (e.g., file sharing, transmission, and operation of communication programs).

Users' Reactions and Comments

Posts that generated the most user reactions (above 100) were related to three thematic categories: original teaching materials, #praiseposts extolling teachers' achievements and their students' work, and those related to the category "Geography Teachers off-duty," which includes informal conversations among teachers. The threshold of 100 reactions was chosen because after observation, it was noted that posts with such levels of activity are rare, indicating their particular uniqueness (there were only 84 such posts during the study period). As for the category of sharing original teaching materials, the highest number of reactions occurred for posts featuring materials that required significant effort or additional skills to create (such as contour maps of continents, topographic maps of soil profiles, elaborate multimedia presentations, or even sketch notes). Posts aimed at showcasing teachers' activities, received with a great enthusiasm, presented projects of their students, account of implementing original lessons, or the outcomes of setting up Geography laboratory. These posts have served as inspiration for other teachers and as a motivation and driving force for the authors of the posts. Examples of such posts include one about the "Edible Flag" project (250 reactions) and one sharing worksheets or showcasing the final outcome of renovating Geography laboratory (275 reactions).

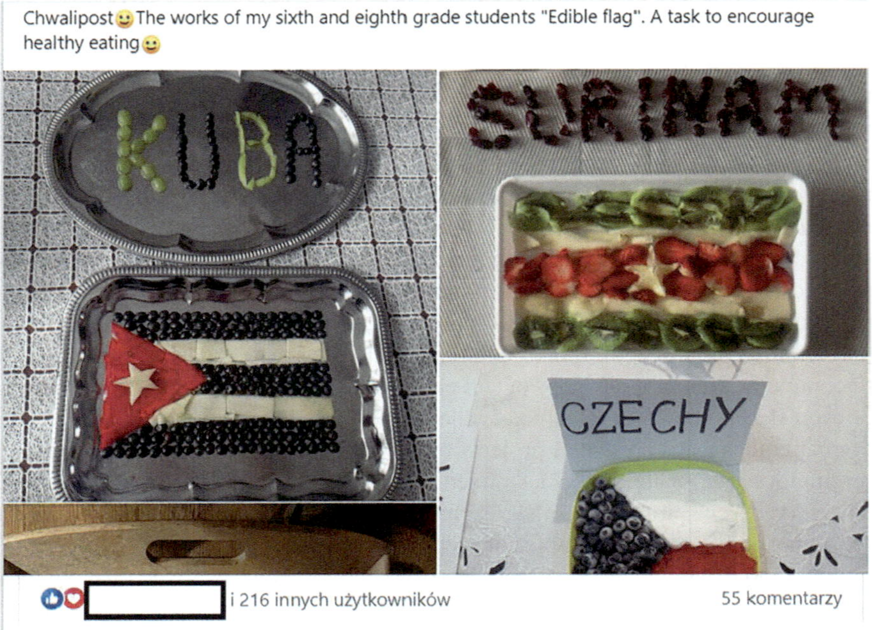

Posts in the category "Geography Teachers off-duty" that garnered many positive reactions and emotions among users often revolved around well-wishes (e.g., for the new hydrological year), funny maps, or creative students' answers during exams. An example could be a post featuring a map depicting the metro network in Poland (despite the fact that there is only a metro system in Warsaw) or wishes from students to teachers celebrating the new hydrological year.

In the context of the thematic scope of the most eagerly commented posts, the discussed topics were diverse but primarily centered around original teaching materials, the teaching profession and pedagogical consultations, remote teaching principles, and requests for inspiration/help in organizing lessons. Depending on the nature of the post, they were commented on more or less frequently. The most popular posts generated close to 200 user comments. The most heated discussions arose under posts related to analyzing assignments and their responses, sharing original materials, pieces of advice for young teachers, or live meetings organized by members of the group. The most commented post concerned an invitation to a seminar for Geography teachers organized by the Commission of Geographic Education of the Polish Geographical Society. It generated as many as 269 comments from Geography teachers eager to participate in the event. It is noticeable that information about meetings and training sessions conducted by scientists educating geographers enjoyed the greatest popularity.

Good morning, on behalf of Prof. , I extend the invitation

for the next meeting in the Seminars for geography teachers series.

Seminar title: **Is geography in primary school too difficult for students and why?**

The seminar will largely be a discussion,

Date: **March 22 (Monday) from 5 p.m. to 7 p.m. (remotely using the Teams platform).**

See more

43 268 comments

The initial stage of remote learning was the time when the most comments were generated under posts. Particularly notable are posts related to thanking fellow group members for their published content, expressing mental support, or even mobilization to action. These posts appeared when the end of the school year or the conclusion of another period of remote teaching was approaching. They have served as the evidence of how valuable and meaningful the group is for teachers.

Thank you very much to all of you for the great materials that I used and which helped me get through the lessons many times in these difficult times.😃. I wish you a good holiday☀☀🏖🌍🌍🌍🏖🌞🌞

64 6 comments

As can be observed, the topics of the most frequently appearing posts align with the most commented and reacted-to entries. However, comments are less popular form of reaction, as half of the published posts were not commented on. The least popular were the announcements about paid conferences aimed at teachers of all subjects, conducted by training companies. Similarly, posts selling original teaching materials did not garner much interest (sometimes there was no reaction to posts on this topic whatsoever). It is apparent that Geography teachers value sharing their original teaching materials and they are engrossed in enhancing their skills through participation in training sessions and conferences dedicated to them.

Critical Event Detection

The technique of post referencing was used to detect critical events. It was assumed that significant moments and events would result in an increased number of forum user posts. Thus, the frequency analysis revealed that during the period from March 1, 2020, to March 31, 2023, nearly 8500 posts were published, and their number was not evenly distributed throughout the study period. On average, 235 posts were published per month during the study period. Months of significantly heightened activity among group members as well as "quieter" periods were observed. These include the beginning of the study period, coinciding with the onset of remote learning in Poland. In March, April, and May 2020, nearly 20% of the posts from the study period were published (1630). The maximum monthly total occurred in May 2020 when over 600 posts were published (a one-time occurrence). Another noteworthy

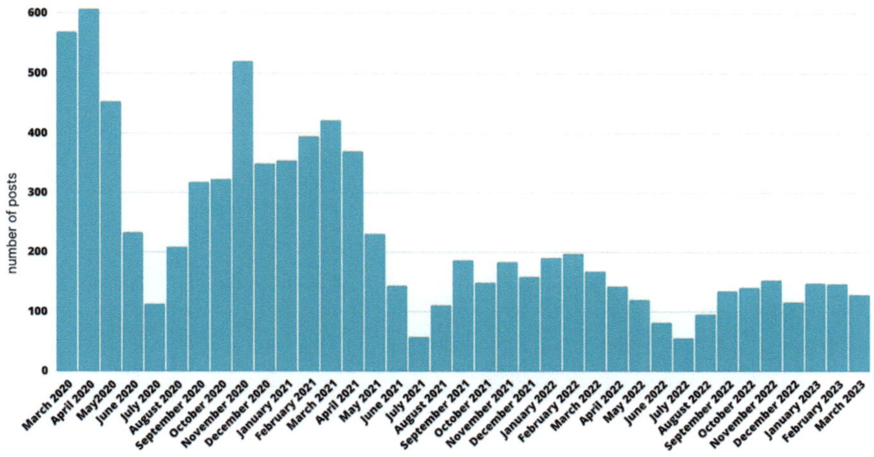

Fig. 4 Total of the published posts on a monthly basis from March, 2020, to March, 2023

month was November 2020, during which approximately 70% more posts were published than in the preceding month (October 2020). An increased activity is associated with the return to remote education for children and youth during that time. In the 2021/22 school year, there was not such a dramatic increase in monthly terms, but in February 2022, the highest number of posts was published. An increased activity was also related to the outbreak of war in Ukraine. Both in monthly and daily terms, it has been observed that despite the occasional deviations, there are cyclic periods of both decreased and increased user activity during the school year. Periods of a reduced user activity occur at the beginning of vacations (July is consistently the month with the lowest total of published posts) and at the end of the calendar year (December). Activity increases from the month preceding the start of the new school year (August in Poland) until the end of the calendar year, followed by a gradual decline until the vacation period (June) (Fig. 4).

In general, three main moments of an increased user group activity, expressed by a significant number of posts on a daily basis, were registered (marked on Fig. 5). The highest number of posts, i.e., 65, was recorded on April 1, 2020. Significant increases in published posts (over 40) also occurred in the period March 18–20, March 24–27, and March 30, 2020. This was related to the introduction of a new form of teaching at that time, namely nationwide remote lessons. The noticeable increase in May 2020 was again related to the effects of the pandemic and information about the extension of remote learning. Another sudden increase was observed on November 10, 2020, the day after the reintroduction of remote learning nationwide due to an increased number of infections. Not only did announcements about the continuation of remote learning trigger an increased user activity. It has been also noticed that the information about returning to stationary lessons stirred emotions among group members. During the analyzed period, user activity decreased with the end of remote learning (the last nationwide remote learning ended on February 27, 2022). There

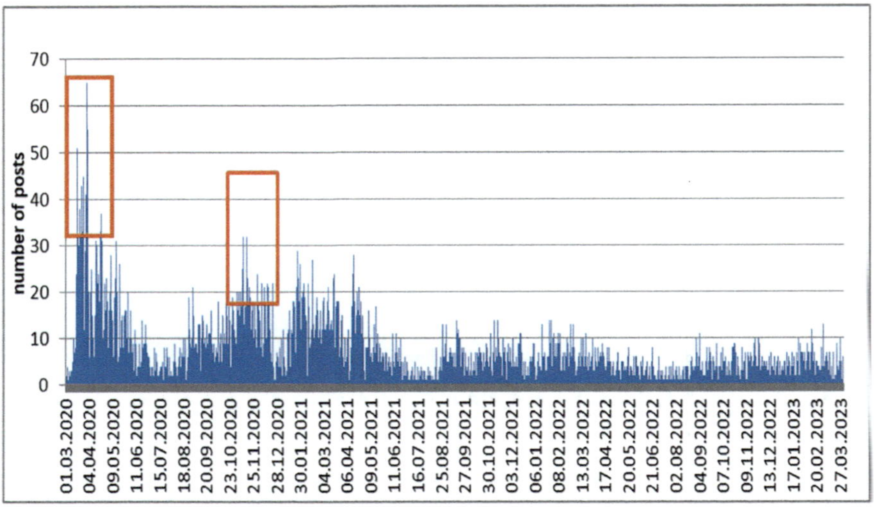

Fig. 5 Total of the published posts on a daily basis from March 1, 2020, to March 31, 2023

were also days during the analyzed period when not a single new post appeared, but this mainly coincided with the summer holidays.

Based on the time series and the content analysis of the posts, the most significant issues that appeared in the entries during the study period were identified, revealing themselves with an increased intensity of added content over time and an increased number of generated comments and reactions. The discussed issues were related to events that were the result of changes in the school environment during the study period. Among them, those that caused the highest user activity were treated as critical events. These were two leading and simultaneously key critical events that had an impact not only on the school environment but on the entire society as well, namely:

(a) the repeated implementation of remote learning;
(b) the outbreak of war in Ukraine.

The period of remote learning was clearly marked on the chart by a significant increase in user activity in terms of the published content on a daily basis. Due to the scale of posts related to remote learning in the earlier period and the overlap of the duration of remote learning with the outbreak of the war in Ukraine, there is no visible increase associated with the war in Ukraine on this chart.

During the analyzed period, there are also complementary critical events that may be significant and important only for the school environment, including teachers, students, or parents. These include changes in the system of teachers' professional advancement and the introduction of pedagogical innovations, subject diagnostics, organizing school trips, equipping subject laboratories, or the situation of teachers in the job market (Fig. 6).

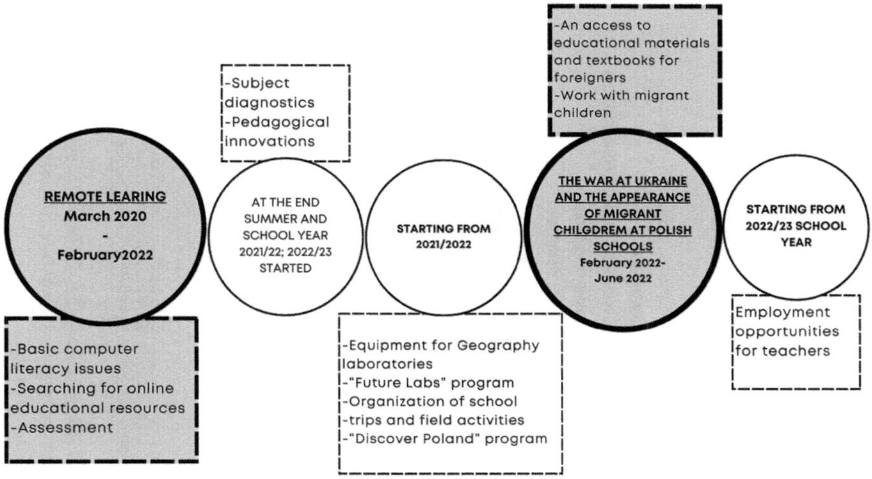

Fig. 6 Key critical events in the years 2020–2023

Based on the time-series analysis, it is observed that the introduction of remote learning in Poland resulted in an increased posting activity in the discussed group. Remote learning was initiated in Poland on March 25, 2020, while the suspension of educational activities occurred on March 11, 2020. These central decisions are illustrated on the chart by the "explosion" of posts on the discussion forum in March and April 2020. A sudden increase in user activity is noticed from March 18, 2020, when the government announcement indicated the introduction of mandatory remote learning. During this time, dozens of posts appeared on the forum daily. During the initial weeks of remote learning, the content of posts solely revolved around the topic of remote learning. Posts related to online educational materials were interspersed with inquiries regarding the process of conducting remote lessons (e.g., communication programs, assessment, conducting exams, and student engagement). The scale of this phenomenon confirms that during the period from March 13, 2020, to April 13, 2020, 60% of the posts related to formalities and the process of remote learning were published from the analyzed period (Fig. 7).

The exact situation occurred at the beginning of November 2020 when nationwide remote learning was introduced once again. Given that remote learning was implemented several times over the course of two school years (also depending on the region), an intensified posting activity occurred each time, but to varying degrees during the analyzed period. The topic of posts related to remote learning remained prevalent among the published posts. This was associated with the possibility of conducting remote individualized education for children whose health condition does not allow for stationary learning (Fig. 6).

In the case of the outbreak of the war in Ukraine, there was no excessively intense increase in user activity. This was due to the fact that the outbreak of the war coincided with the return to stationary education in Poland. However, upon a closer analysis of

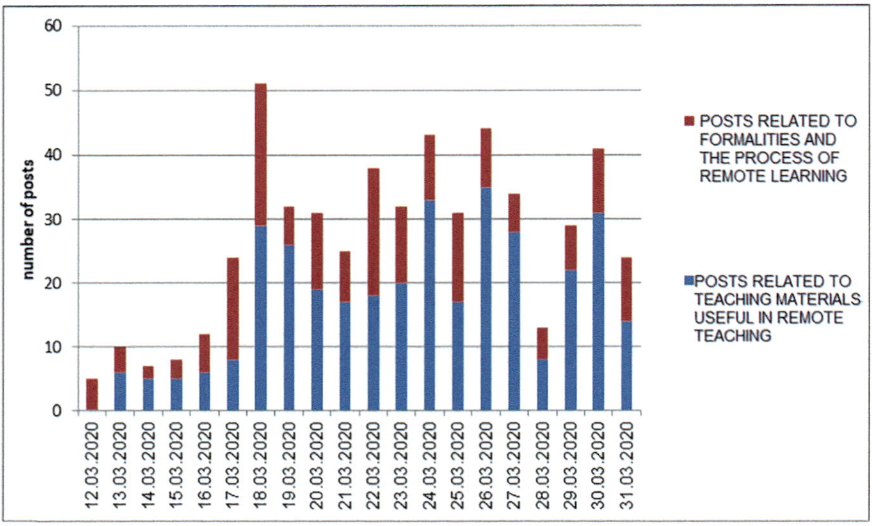

Fig. 7 Topics of the published posts in the period from March 12, 2020, to March 31, 2020

the content of the posts, it can be noticed that the war and the mass influx of refugee students into the Polish education system reflected in the topics of the posts. During the period from February 24, 2022, to March 14, 2022, i.e., during the first three weeks since the outbreak of the war, 60% of the posts published in "Geography Teachers" group concerned the integration of Ukrainian students into the Polish school system (Fig. 8).

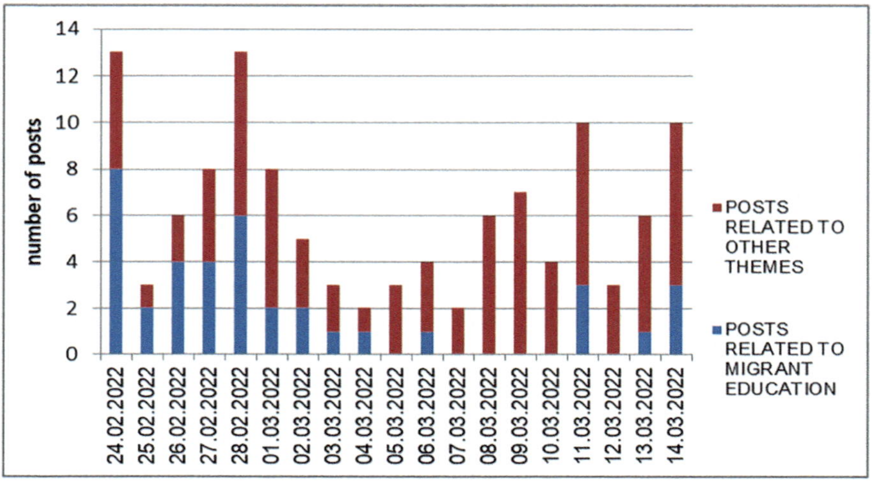

Fig. 8 Topics of the published post from February 24, 2022, to March 14, 2022

These posts were related to the search for educational materials in the Ukrainian language or consultations regarding working with students with migration experience. Examples of such posts include materials published in the Ukrainian language.

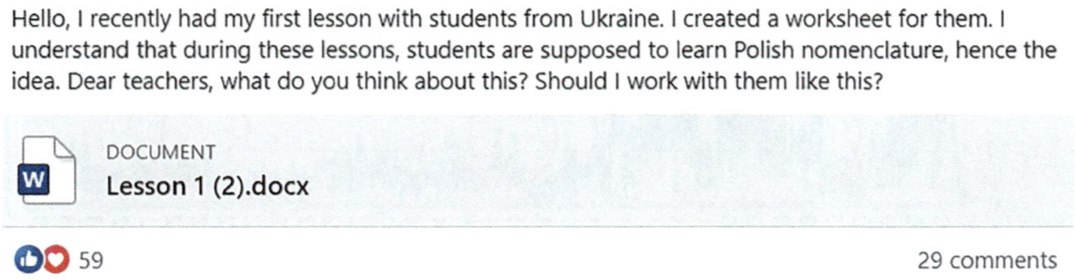
Hello, I recently had my first lesson with students from Ukraine. I created a worksheet for them. I understand that during these lessons, students are supposed to learn Polish nomenclature, hence the idea. Dear teachers, what do you think about this? Should I work with them like this?

DOCUMENT
Lesson 1 (2).docx

59
29 comments

Both events have had a significant impact on the way teachers work, students learn, and educational institutions operate in general. These events resulted from many changes that occurred and will be visible for a long time in the educational environment. Observing the emerging posts allowed us to notice changes in the type of the published content. They became more multimedia and universal, suitable for both stationary and remote education, as well as easy to translate into foreign languages.

During the analyzed period, three other moments were noted that, while not as decisive as the discussed events, clearly engaged users as well. Therefore, these events can be considered critical incidents of a lesser scale and intensity in their impact on the work of Geography teachers. These episodes appeared at the end of the summer holidays and the initial weeks of the new school year (August, September). The start of a new school year usually brings about amendments to regulations or the announcement of new projects implemented by schools. Additionally, a certain cyclical nature of the appearance of specific content themes in certain periods of the school year was observed, indicating that the new challenges posed to teachers may also be perceived as critical by some of them (Fig. 9).

The turning point of each August and September during the analyzed period marks an increase in number of monothematic posts. Among the typical content related to teaching materials, prior to and shortly after the start of the 2021/22 and 2022/23 school years, there was an observed heightened intensity of inquiries regarding pedagogical innovations, as well as educational diagnostics. This was associated with changes in the system of teachers' professional advancement, according to which one of the duties was to introduce innovations to meet the requirements for advancement. Posts published during this time focused on seeking inspiration for creating innovations or school projects (Fig. 10).

Because of remote learning, which was present throughout the analyzed period, an increasing number of schools opted to conduct educational diagnostics, i.e., tests verifying the level of knowledge acquired in the previous school year. This form of knowledge assessment was not widespread in Poland, but remote learning exerted pressure to evaluate the outcomes of distance education (an example).

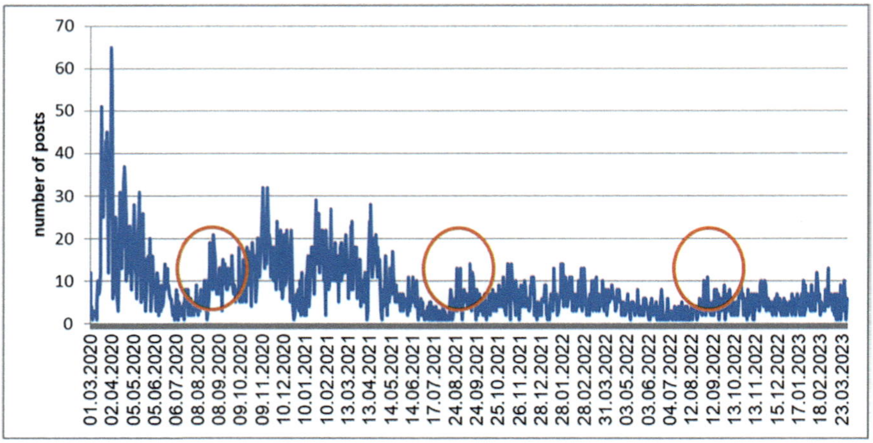

Fig. 9 Total number of posts published daily from March 1, 2020, to March 31, 2023

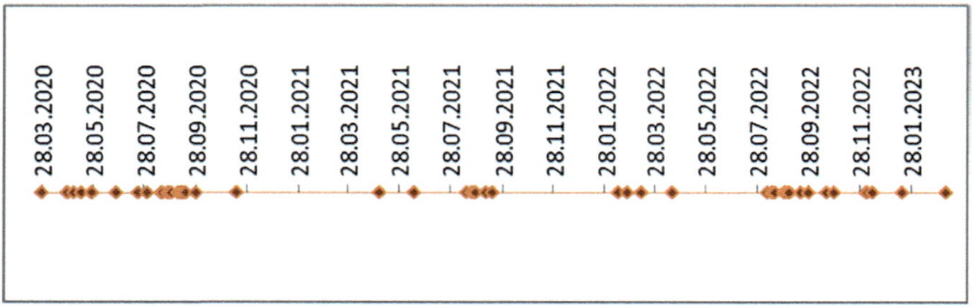

Fig. 10 Published posts concerning pedagogical innovations and educational diagnostics (March 1, 2020–March 31, 2023)

#diagnosis
Good evening. The director asked for diagnostic tests to find out the deficiencies from last year. I will be teaching 6th grade (I have just finished my postgraduate studies). Where can I find appropriate tests? Could someone share such tests? I will be very grateful for your help🙏🙏🙏

11 comments

The 2021/22 school year marked the continuation of the Educational Reserve program and the announcement of new educational projects, significantly related to Geography, thus creating a new trend in the themes of published posts. The Ministry of Education and Science introduced two significant programs, "Discover Poland" and "Future Labs." Consequently, discussions on the geographers forum revolved around the organization of school trips and the formalities associated with them. School trips and fieldwork are essential elements of geographic education. For Geography teachers, this is particularly important, as they are usually the organizers and initiators of field trips or geographical excursions. Posts regarding trips and fieldwork

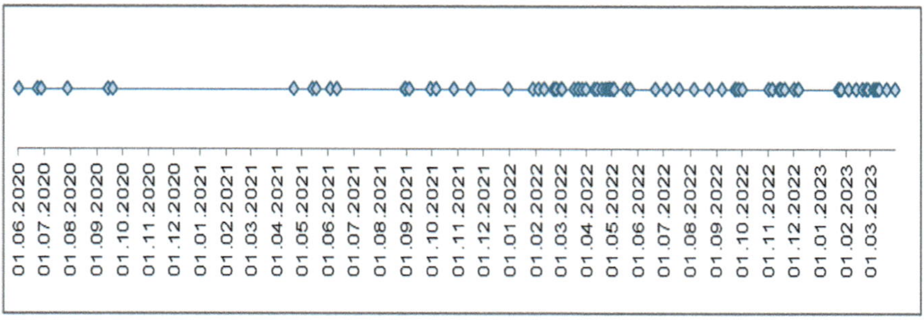

Fig. 11 Published posts concerning the topic of field trips and outdoor activities between March 1, 2020, and March 31, 2023

included inquiries about legal formalities, recommendations for accommodation, guides, or ideas for trip destinations. A post may be an example here:

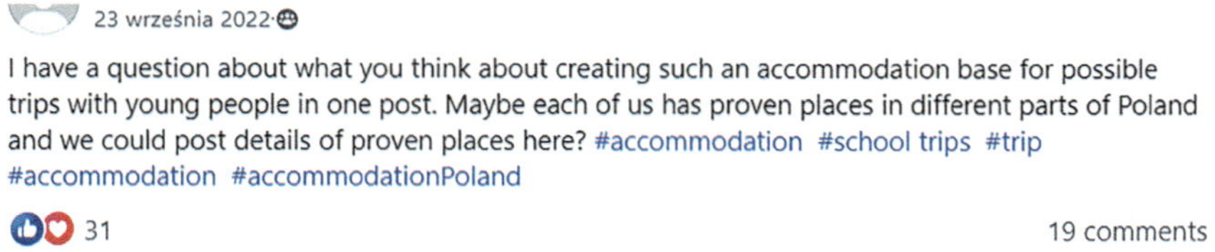
23 września 2022·

I have a question about what you think about creating such an accommodation base for possible trips with young people in one post. Maybe each of us has proven places in different parts of Poland and we could post details of proven places here? #accommodation #school trips #trip #accommodation #accommodationPoland

31 19 comments

Some of the posts were a result of decisions made by the Ministry of Education and Science, which announced the "Discover Poland" school trips program starting from the school year. The program aimed to subsidize school trips by the Ministry to increase the popularity of domestic trips to places associated with the tradition, culture, and history of a given region.

Since the program's announcement, the topics of the posts have been expanded to include issues related to the procedure for applying for this funding and ideas for implementing these activities (Fig. 11).

Schools have the opportunity to apply for funding for renovations and equipping laboratories. Teachers oversee subject-specific laboratories, identify the need for equipment, materials, and teaching aids, and submit requests. In connection with participation in these programs, teachers sought reliable and useful teaching materials such as maps, models, globes, and diagrams. Some posts regarding this topic focused on seeking both inspiration and assistance related to designing the decor of Geography laboratory. An increase in inquiries related to laboratories was observed from 2021 onward when the "Future Labs" program was announced in Poland, aimed at equipping schools with the state-of-the-art equipment and laboratories.

Subsequently, a pool of posts illustrating problems related to the shortage of teachers in Polish schools was noted. The number of posts in the group regarding job offers for Geography teachers increased. During the analyzed period, 88 job offers for Geography teachers were announced in the group, covering various regions of Poland. In terms of vacant positions for Geography teachers, the current school year

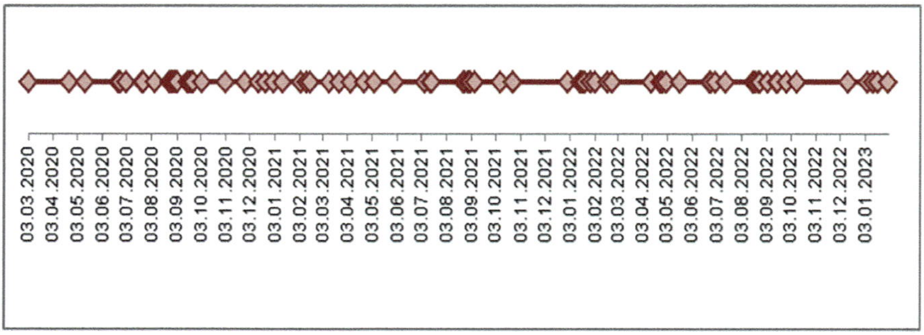

Fig. 12 Published posts concerning seeking Geography teachers for employment (March 1, 2020–March 31, 2023)

(2023/24) is record-breaking, but the deepening of this phenomenon has been visible since the beginning of the analyzed period, as evidenced by an increasing number of job offers posted in "Geography Teachers" group. This could be the evidence that low wages, professional burnout, and the lack of societal respect for the teaching profession are causing a personnel crisis in the Polish education (Fig. 12).

Summary

The research subject, a virtual discussion group comprising Polish Geography teachers, along with the application of a classification key and the content analysis method, allowed for the identification of primary and secondary critical events that occurred in the school environment during the studied period. The analysis of the research results indicates that during the studied period, two critical events of particular significance and five of lesser importance occurred. The most important events classified were the outbreak of war in Ukraine, which resulted in migration movements and the presence of the Ukrainian children and youth at Polish schools, and a large-scale remote learning. Both of these events fit the characteristics of critical events which, as previously defined, are sudden and unexpected occurrences that have a significant impact on society.

Among the lower-ranking events, the introduction of projects and pedagogical innovations, conducting subject diagnostics, organizing field trips and school excursions, including the use of projects related to school trips (such as "Discover Poland" program), were included. The implementation of projects funded by external institutions aimed at improving the equipment of subject laboratories (e.g., "Future Labs") as well as the increasingly visible societal problem associated with the teacher shortage crisis were also classified as lower-ranking critical events.

There is a noticeable recurrence of topics that were consistently discussed in "Geography Teachers" group throughout the school year. These included posts concerning lesson inspiration (methods, resources, and scenarios), designing open

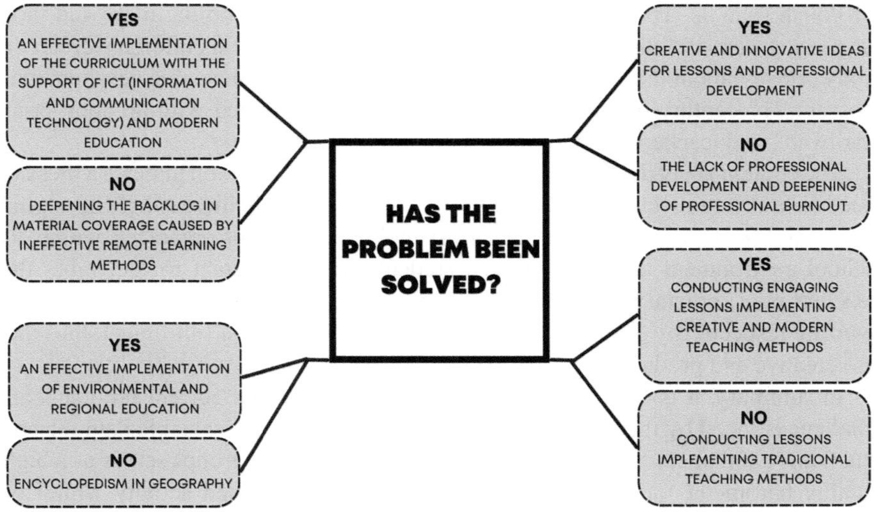

Fig. 13 Sample effects of critical events with consideration of teachers' actions and educational authorities' direction

lessons, and acquiring access to teaching materials (diagrams, graphics, tasks, quizzes, and online maps). The recurring issues could serve as a harbinger of changes for educational authorities, as these challenges and problem areas occur at particular times during the school year. Skillfully addressing a given challenge ensures that the quality of education provided by a teacher or a school remains high. However, if the system fails to address the problem adequately, the recurring issue will deepen and worsen, leading to a decline in the quality of education.

Therefore, it is important to draw conclusions and consider how to address emerging problems at an early stage. Analyzing and predicting the consequences of solving the problem or the lack thereof will help improve the quality of education and the working conditions of teachers. In the opinion of the author of the article, group discussion forums serve as a seismograph for the tasks and challenges faced by teachers and students during the school year. Monitoring the intensity of emerging posts will help minimize the negative experiences associated with them (Fig. 13).

Conclusion

The past three years have been particularly challenging for Polish teachers in terms of facing various challenges and uncertainties. By means of the empirical analysis of forum posts, it is evident that there is an insufficient systemic methodological support for teachers in Poland in terms of teaching methods and techniques. This is also associated with visible difficulties among teachers in using ICT tools both during remote and stationary education. This is an important issue concerning the future of Polish schools. The children enrolling at schools are becoming more and more increasing part of a digital society. It is important not to stifle children's creativity and curiosity about the world and their surroundings. To achieve this, teachers must possess 21st-century digital skills. Schools must be equipped with an appropriate hardware, and teachers must know how to use them effectively.

The content analysis of forum posts and thematic confrontation indicate a pressing need for a strong psychological support for teachers, who are increasingly struggling with burnout. Unfortunately, so far, a psychological support organized within the school environment has only targeted students. It is important to remember that teachers play a crucial role in shaping and educating the younger generation. Teachers who are respected by students and parents feel satisfaction and fulfillment, thus they are creative and productive in their work.

In summary, it is observed that discussion groups reflect current problems and challenges faced by their users and provide researchers with valuable data for both qualitative and quantitative analysis. Analysis of discussion groups serves as a high-quality barometer for critical events that lead to an increased activity within the groups. However, most importantly, group analysis can help rebuild societal respect for teachers and foster an open, creative school environment fit for the future.

References

Al-Douri J (2021) Grupy zamknięte na portalach społecznościowych inspiracją dla nauczycieli edukacji plastycznej. Świat plastyki w wirtualnej rzeczywistości. Pedagogika Przedszkolna i Wczesnoszkolna 9(1):101–121. https://doi.org/10.4467/23537159PPW.21.009.13993

Athanassopoulos S, Lavidas K, Komis V (2021) Student Facebook group as a discussion platform for academic purposes. In: ICERI2021 proceedings, pp 8071–8077

Bąk A (2016) Serwisy społecznościowe—efekt Facebooka i nie tylko, w: „Media i Społeczeństwo". Akademia Techniczno-Humanistyczna, Bielsko-Biała 6:134–146

Bergviken-Rensfeldt A, Hillman T, Selwyn N (2018) Teachers 'liking' their work? Exploring the realities of teacher Facebook groups. Br Edu Res J 44(5):230–250

Bowman N, Akcaoglu M (2014) "I see smart people!": using Facebook to supplement cognitive and affective learning in the university mass lecture. Internet High Educ 23:1–8

Chęcińska-Kopiec A, Ochoa Siguencia L (2020) MEDIA SPOŁECZNOŚCIOWE I PLATFORMY E-LEARNINGOWE W EDUKACJI, Częstochowa: Instytut Badań i Innowacji w Edukacji [INBIE]

Chromińska A, Raciborski F (2019) Fora internetowe poświęcone atopowemu zapaleniu skóry—wyniki analizy treści. Alergia Astma Immunologia. 24(2):68–74

Cisek S (2017) Diagnostyka informacyjna w miejscu pracy- technika incydentów krytycznych, In: Sapa R (eds) Diagnostyka w zarządzaniu badaniem: perspektywa informatologiczna, Biblioteka Jagiellońska, Kraków, pp 129–141

Dołowy-Rybińska N (2017) „Nikt za nas tego nie zrobi" praktyki językowe i kulturowe młodych aktywistów mniejszości językowych Europy. Wydawnictwo Naukowe Uniwersytetu Mikołaja Kopernika, Toruń

Fabjaniak-Czerniak K (2012) Internetowe media społecznościowe jako narzędzie public relations. In: Kubiak K (ed) Zarządzanie w sytuacjach kryzysowych niepewności. Wyższa Szkoła Promocji, Warszawa, pp 173–194

Forst K (2017) Facebook we współczesnym świecie komunikacji społecznej. In: Doktorowicz K (eds) Media społecznościowe: dialog w cyberprzestrzeni. T. 2, Katowice: Wydawnictwo Uniwersytetu Śląskiego, pp 37–55

Grębosz M, Siuda D, Szymański G (2016) Social media marketing. Wydawnictwo Politechniki Łódzkiej, Łódź

Iwanowska M (2016) Zrozumieć media (społecznościowe). Perspektywa psychologiczna. In: Łosiewicz M, Ryłko-Kurpiewska A (eds) MEDIA BIZNES KULTURA. Rzeczywistość medialna. Formy, problemy, aspiracje, Gdynia-Kinvara Co. Galway: Novae Res-True Sign, 3, pp 169–185

Janc K, Jurkowski W (2020) Media społecznościowe w czasie pandemii–przykład profili wrocławskich rad osiedla na Facebooku. Rozwój Regionalny i Polityka Regionalna (51):9–24. https://doi.org/10.14746/rrpr.2020.51.03

Jaskiernia A (2021) Pandemia COVID-19 i „nowa normalność" w systemie komunikowania. Wyzwania regulacyjne epoki cyfry. Media Biznes Kultura 2(11):9–21

Juza M (2020) Komunikacja językowa w Internecie: od języka Internetu do pragmatyzmu komunikacji. In: Kępa-Figura D, Naruszewicz-Duchlińska A (eds) Polszczyzna w dobie cyfryzacji. Czasopisma i Monografie PAN, Warszawa, pp 17–28

Klepka R (2016) Analiza zawartości mediów: dlaczego i do czego można ją wykorzystać… w nauce o bezpieczeństwie i politologii?, Kraków: Annales Universitatis Paedagogicae Cracoviensis. Studia De Securitate Et Educatione Civili 224(6):32–41

Kosik E (2017) Bibliofil w Internecie, czyli grupy poświęcone tematyce książek w serwisie Facebook. Toruń: Toruńskie Studia Bibliologiczne 10(2):241–159. https://doi.org/10.12775/TSB.2017.025

Krawczyk-Bocian A (2013) Doświadczanie zdarzeń krytycznych. Narracje biograficzne dorosłych dzieci alkoholików, Bydgoszcz : Uniwersytet Kazimierza Wielkiego

Kubica W (2022) Praca w szkole online—czyli o perspektywie nauczycielek i nauczycieli na podstawie analizy treści postów zamieszczonych w grupie społecznościowej serwisu Facebook „Ja, nauczyciel". Kraków: Youth in Central and Eastern Europe 9(13):67–74. https://doi.org/10.24917/ycee.9559

Kućko W (2019) Etyczne aspekty mediów społecznościowych oraz ich wpływ na zachowanie i sposób myślenia na początku XXI w. Społeczeństwo i Polityka 4(61):43–65. https://doi.org/doi.org/10.34765/sp.0419.a03

Kulesza-Gulczyńska B (2016) Facecje na Facebooku – studium przypadku. Forum Poetyki 4–5:70–81. https://doi.org/10.14746/fp.2016.4-5.26717

Kuryś-Szyncel K, Błachnio A (2020) Krytyczne wydarzenia życiowe i ich znaczenie w biografiach seniorów. Nauki o Wychowaniu Studia Interdyscyplinarne. 10(1):114–136. https://doi.org/10.18778/2450-4491.10.09

Leung XY (2012) The marketing effectiveness of hotel facebook pages: from perspectives of customers and messages, Las Vegas: UNLV Theses, Dissertations, Professional Papers, and Capstones, 1680. https://doi.org/10.34917/4332661

Levickaite R (2010) Generations X, Y, Z: how social networks form the concept of the world without borders (the case of Lithuania). Vilnus: Limes Cultural Regionalistics 3(2):170–183

Lupa-Wójcik I (2020) Groups on Facebook as a marketing tool. Marketing Instytucji Naukowych i Badawczych 37(3):17–32

Łosiewicz M (2013) Wykorzystanie mediów społecznościowych w komunikacji biznesowej. In: Kaczmarczyk M (ed) Faces of social communication: around the issues of marketing and public relations in contemporary society, Sosnowiec: Oficyna Wydawnicza „Humanitas". Verbum, Praha, pp 135–144

Majewska J (2021) Wpływ pandemii COVID-19 na znaczenie mediów społecznościowych w Tajlandii. Ekonomia XXI Wieku 24:101–115

Marek E (2017) Facebook źródłem doświadczeń, inspiracji i rozwoju nauczycieli z grupy "Nauczyciele wczesnoszkolni." Pedagogika Przedszkolna i Wczesnoszkolna. 5(1):89–110

Meta Reports Fourth Quarter and Full Year 2023 Results; Initiates Quarterly Dividend. https://investor.fb.com/investor-news/press-release-details/2024/Meta-Reports-Fourth-Quarter-and-Full-Year-2023-Results-Initiates-Quarterly-Dividend/default.aspx
Naruszewicz-Duchlińska A (2005) Wirtualny Hyde Park—kilka uwag o komentarzach internetowych. Media—Kultura—Komunikacja Społeczna 1:249–253
Paczka E (2020) Zmiana zachowań rynkowych pokolenia Z. Ekonomia—Wroclaw Econ Rev 26(1):21–34
Perz P (2022) Korzystanie z portali społecznościowych przez badane grupy dzieci, rodziców i nauczycieli. Zeszyty Naukowe Wyższej Szkoły Gospodarki, Seria Edukacja—Rodzina—Społeczeństwo 7:305–326
Pietrzak-Zawadka J, Lewon R (2018) Media społecznościowe i ich znaczenie w edukacji przyrodniczej. Studia i Materiały Centrum Edukacji Przyrodniczo-Leśnej 20(1[55]):86–92
Płuciennik J (2014) Pomiar aktywności w mediach społecznościowych—wyzwanie współczesnych badań marketingowych. Marketing i Rynek 3(CD):145–158
Przybysz J, Pioterek P (2015) Media społecznościowe w służbie bibliotek In: Brzezińska-Stec H, Żochowska J (eds) Biblioteki bez użytkowników...? Diagnoza problemu, Białystok: Wydawnictwo Uniwersytetu w Białymstoku, pp 339–356
Razmerita L, Kirchner K (2014) Social media collaboration in the classroom: a study of group collaboration. In: Baloian N, Burstein F, Ogata H, Santoro F, Zurita G (eds) Collaboration and technology. CRIWG 2014. Lecture Notes in Computer Science, 8658. Springer, Cham. https://doi.org/10.1007/978-3-319-10166-8_25
Rynkiewicz W (2011) Facebook jako ważny środek komunikacji studentów z uczelniami wyższymi Wrocławia. Zeszyty Naukowe/uniwersytet Ekonomiczny w Poznaniu 209:107–117
Szczepaniak K (2012) Zastosowanie analizy treści w badaniach artykułów prasowych—refleksje metodologiczne, Acta Universitatis Lodziensis. Folia Sociologica 42:83–112
Thelwall M (2008) Social networks, gender, and friending: An analysis of MySpace member profiles. J Am Soc Inform Sci Technol 59(8):1321–1330. https://doi.org/10.1002/asi.20835
Thelwall M, Wilkinson D, Uppal S (2009) Data mining emotion in social network communication: gender differences in MySpace. J Am Soc Inform Sci Technol 61(1):190–199. https://doi.org/10.1002/asi.21180
Tripp D (1996) Zdarzenia krytyczne w nauczaniu. Kształtowanie profesjonalnego osądu. Wydawnictwo Wydawnictwa Szkolne i Pedagogiczne, Warszawa
Wąsowicz-Zaborek E (2020) Analiza treści generowanych przez użytkowników Internetu jako podstawa decyzji marketingowych w małym przedsiębiorstwie. In: Bohdanowicz L, Dziurski P (eds) Innowacje i marketing we współczesnych organizacjach: wybrane zagadnienia. Oficyna Wydawnicza SGH—Szkoła Główna Handlowa w Warszawie, Warszawa, pp 281–301
Wolny W (2017) Wielowymiarowa analiza mediów społecznościowych. Ekonomiczne Problemy Usług 126(2):305–313
Wójtowicz M (2019) Facebook jako współczesne narzędzie komunikacji i kreatywności językowej. In: Burska K, Cieśla B (eds) Kreatywność językowa w komunikacji internetowej. Wydawnictwo Uniwersytetu Łódzkiego, Łódź, pp 151–162
Zarzycka G (2010) Incydenty krytyczne, czyli najtrudniejsze doświadczenia słuchaczy Studium Języka Polskiego dla Cudzoziemców UŁ. Acta Universitatis Lodziensis Kształcenie Polonistyczne Cudzoziemców 17:63–76

Sustainability as a Future Challenge

Post-growth as a Topic in Geography Education—Suggestions from Secondary School Students in Germany in the Context of Education for Sustainable Development

Lara Brede, Andreas Eberth, and Christiane Meyer

Abstract The term "post-growth" covers various approaches that decouple from the current economic growth paradigm and strive for human well-being and prosperity beyond economic growth as a goal. As young people's involvement is central in discussions about shaping our future through developing capacities as change agents, this topic offers considerable potential for transformative literacy in the context of Education for Sustainable Development (ESD). In a research project at Leibniz University Hannover, seven groups of 25 young people from upper secondary schools participated in three online meetings to reflect on their ideas about post-growth; the first and last sets of meetings were conducted as focus groups. The first set focused on different points of criticism of the economic growth paradigm and alternative approaches, after which the participants selected a change agent that was particularly interesting to them. In the second set of meetings, the participants interviewed the change agents about their contributions to the co-creation of a post-growth society. This chapter focuses on the final set of meetings, which addressed participant insights from the expert interviews and the proposals of the young people on how to implement this topic in geography lessons. Our results emphasise the need for integrating change agents into school education and a stronger focus on post-growth-related topics.

This research was funded by the Ministry of Science and Culture of Lower Saxony, grant number 74ZN1809.

L. Brede (✉) · C. Meyer
Institute of Science Education, Geography Education Section, Leibniz University, Hannover, Hanover, Germany
e-mail: brede@idn.uni-hannover.de

C. Meyer
e-mail: meyer@idn.uni-hannover.de

A. Eberth
Professorship for Geography with a focus on Education for Sustainable Development, University of Passau, Passau, Germany
e-mail: andreas.eberth@uni-passau.de

© The Author(s), under exclusive license to Springer Nature Switzerland AG 2025

D. Piróg and A. Świętek (eds.), *Future-Ready Geography*, Key Challenges in Geography, https://doi.org/10.1007/978-3-031-81268-2_7

Keywords Post-growth · Education for Sustainable Development (ESD) · Transformative literacy · Change agents

Introduction

The 2030 Agenda for Sustainable Development (UN 2015) defines and outlines the 17 Sustainable Development Goals (SDGs). SDG 8 focuses on "[promoting] sustained, inclusive and sustainable economic growth, full and productive employment, and decent work for all" (UN 2015, 14). Therefore, although researchers like Maja Göpel (2016, 3) state that "the most critical aspect for turning the wheel towards fulfilling the SDGs is changing the economic paradigm", economic growth is still prioritised as the guiding principle for our economic system in SDG 8—and by extension, the UN's 2030 Agenda.

After critically reflecting on SDG 8, we outline some post-growth approaches in this chapter. Young people are pivotal in shaping our future, and their empowerment is essential for Education for Sustainable Development (ESD) (UNESCO 2020, 32). This context motivated our focus on young people's geographies (Lambert 2009). Ensuring the involvement of young people is possible by integrating post-growth topics in (geography) education, which is why we discuss the implications for ESD in school education and the current status quo of post-growth in secondary schools in Lower Saxony, Germany. To integrate young people as active participants, we investigated the perspectives of young people in Germany on economic growth and post-growth, as well as the involvement of change agents in our research project. The following sections detail the results of this chapter's analysis and discussion, focusing particularly on the participants' ideas for addressing post-growth in geography lessons.

Theoretical Framing

Critical Reflections on Sustainable Development Goal 8

The five guiding principles of the 2030 Agenda are people, planet, prosperity, people, and partnership (UN 2015). In that respect, prosperity is often considered as being directly connected to economic growth, which is understood as "an increase in the goods and services produced by an economy in a given period, typically a year", which is measured and tracked as a country's "increase in Gross Domestic Product (GDP)" (Victor 2015, 109). Due to the antagonism between neoliberal capitalism and ecology, this economic growth paradigm, even as the guiding principle of our economic system, is increasingly criticised within geographical and sustainability debates (Haraway 2016; Harvey 2020; Kallis 2018; Klein 2014; Potter et al. 2012) and the negative impacts on social equality as well as on the environment and climate change (Kallis et al. 2018). Consequently, researchers like Tim Jackson (2009), author of *Prosperity Without Growth* (originally released as a report by the Sustainable Development Commission), advocate for system change. To this end, new business models and practices are required (Peck 2009; Paech 2012). Nonetheless, SDG 8 only offers suggestions for integrating a more sustainable frame into the growth paradigm without providing specific proposals for implementation (Eberth et al. 2023).

Concepts for a Post-growth Era

The term post-growth encompasses various alternative approaches characterised by detachment from the current economic system focused on growth. In these approaches, long-term economic growth via technological innovations or improved efficiency is seen as impossible (Schulz et al. 2021).

> In essence, it is about adjusting understandings of growth and re-evaluating it, examining the long-term meaningfulness of certain developments, and, if necessary, looking for possible alternatives within free social conditions. Meaningfulness refers here not only to the environment but also to individual and social needs, i.e. focus [on] the common good rather than individual profitability. (ibid., 20)

Contrary to post-growth, the term "degrowth" is detached from the growth discourse and defined as "a process of political and social transformation that reduces a society's throughput while improving the quality of life" (Kallis et al. 2018, 292).

In the field of post-growth economic research, approaches can occur at different levels, as illustrated in Fig. 1. These levels—(1) system knowledge and (2) target knowledge, each transmitted via models and theories, and (3) transformation knowledge, which is conveyed through practices of action—are based on the concept of transformative literacy, which is detailed in Sect. "Transformative Literacy in the Context of ESD" (Eberth et al. 2023; Singer-Brodowski and Schneidewind 2014). Aside from the mentioned initiatives, young people's involvement is central in discussions about shaping our future (UNESCO 2020).

When examining post-growth economic approaches, geographical aspects to transformation research, including the social production of space and spatial relations, must be considered. In this context, the question on how different civil society initiatives are scaled to overcome structural disparities between the local and global (Marston et al. 2005; Massey 2008) and the differentiation between bottom–up and top–down implementation strategies of social change (Gallo-Cruz 2017) are much-debated issues. Furthermore, "governance and planning-related contributions raise questions about the reference areas, spaces of action and territoriality of post-growth processes" (Schulz et al. 2021, 21).

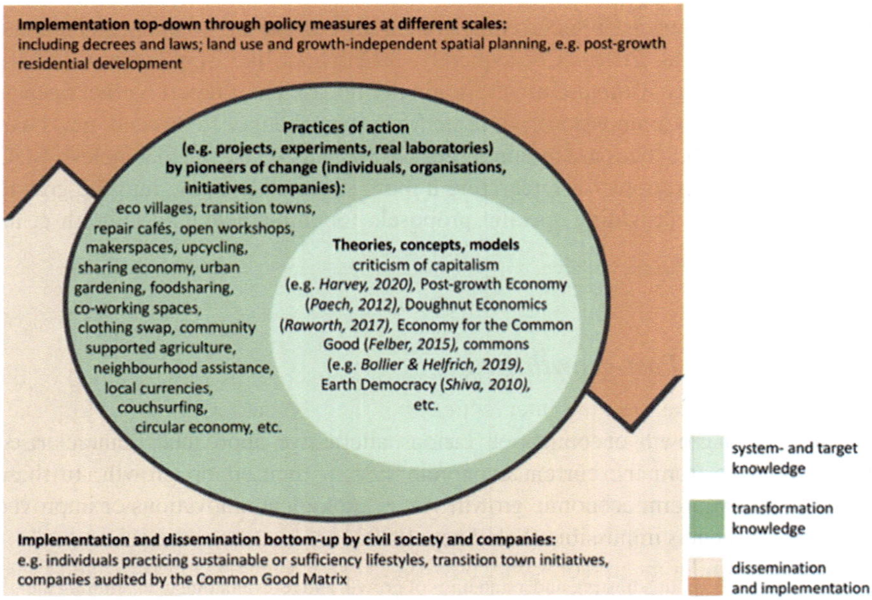

Fig. 1 Approaches in the field of post-growth (Eberth et al. 2023, 159)

Young People's Geographies

Young people are pivotal in developing ideas and a post-growth society and economy (UNESCO 2020). As geographical perspectives of space and spatial relations are vitally important in the context of post-growth economics, geographical research concerning young people assumes that their experiences of space are relevant and differ from those of adults (Freeman and Tranter 2015). Young people's geographies thus play a specific role in this process (Lambert 2009).

Concurrently, young people are more widely considered to be active agents within their spaces than ever before (Aitken 2019). Current research on young people's geographies focuses on, among other topics, consumerism, the increased use of technology, and life pace (Freeman and Tranter 2015).

ESD regards young people as active participants who must be encouraged "to explore values which could provide an alternative to consumer societies, such as sufficiency, fairness and solidarity" (UNESCO 2020, 58). For ESD learning to succeed, a new understanding of schools and worthwhile skills must be achieved. As young people are critical agents in creating a post-growth society and economy and shaping our future, they should be recognised as change agents (UNESCO 2020).

ESD for a System Change Towards Post-growth

UNESCO's programme *ESD for 2030* assigns ESD a central role in fulfilling the 17 SDGs. It demands that ESD focusses on a significant transformation for which structural changes are urgently needed. Here, the relationship between economic growth and sustainable development is a decisive factor. The programme also calls for systemic economic change, for which ESD is mandatory to provide alternatives to our consumer society and must "affect the unsustainable production patterns of current economic structures more directly", such that "people [must] be empowered to engage directly in the political process and advocate, for example, for appropriate environmental regulations for businesses" (UNESCO 2020, 58). Transformative action and change agents are each crucial in achieving these efforts (UNESCO 2020).

Transformative Literacy in the Context of ESD

This study adopts the definition of transformative literacy as the ability to "adequately understand the multidimensionality of transformation processes and to integrate one's own actions into transformation processes" (Schneidewind 2013, 83 in WBGU 2016, 425). This form of literacy consists of system knowledge, which refers to the capacity to analyse problems; target knowledge, which refers to the capacity to envision future paths; and transformation knowledge, which refers to the understanding of different conditions needed to realise these visions. Accordingly, transformative literacy corresponds with ESD (Singer-Brodowski and Schneidewind 2014; Meyer and Höbermann 2021). The transformational education process "conceives the interaction between subject and world as the subject's response to new challenges posed by its environment" (Koller 2020, 635) and thus involves transformative literacy. This can be seen as an orientation for transformative education (Meyer and Höbermann 2021). The concept of transformative literacy was used as the framework for our research design, which is presented in Chapter Three.

For this study, we adopt a definition of transformative learning as a profound shift that "involves our understanding of ourselves and our self-location; our relationships with other humans and with the natural world" and "our understanding of power relations in interlocking structures of class, race and gender, our body awarenesses, our visions of alternative approaches to living, and our sense of possibilities for social justice, peace and personal joy" (O'Sullivan 2012, 164). It requires transformative education and an "action-oriented, transformative pedagogy [that] supports self-directed learning, participation and collaboration, problem-orientation, interdisciplinarity and transdisciplinarity, and the linking of formal and informal learning" (UNESCO 2017, 7). Transformative action is a key reflection of ESD and can be triggered by a disruption when people diverge from their usual way of thinking or acting. For system change to occur, individual-level transitions—transformative actions—are indispensable: "ESD is [fundamental] to provide [individuals] with critical thinking skills [that enable them] to reflect on individual values, attitudes, and behaviours as well as lifestyle choices" (UNESCO 2020, 58).

Change Agents as Role Models

Learning institutions, especially schools, need to change to encourage young people to take transformative action and become change agents themselves. In this context, essential factors are opportunities that offer or have "critical inquiry, exposure to realities, relevance to our own lives and the presence of influential peers, mentors or role models, as well as tipping points" (UNESCO 2020, 57). The importance of non-formal education within a community and informal learning outside of school contexts with external experts is thus strengthened (UNESCO 2020).

Accordingly, aside from enabling learners to become change agents themselves, it is necessary to include change agents—sometimes referred to as "pioneers of change"—in school education (Eberth et al. 2023, 159). These can be individuals, groups, initiatives, or companies who "propagate innovation by questioning 'business as usual' policies and creating alternative practices, thereby challenging the established world views and paths, attitudinal and behavioural patterns, as well as providing others who think as they do (followers, early adopters) with a constant motivation for self-sustaining change" (WBGU 2011, 243). Therefore, they function as role models and initiate a local change "from below". Change agents are active in several fields, such as environmental initiatives, social movements, governmental and non-governmental organisations, science, architecture, and urban planning (WBGU 2011).

Status Quo of Post-Growth in Secondary Schools of Lower Saxony, Germany

As young people should be integrated as change agents, school education plays a pivotal role in teaching skills and building awareness, enabling them to become active participants in shaping our future (UNESCO 2020). To investigate if and in which way school education already contributes to this, we must examine the current status quo in German secondary schools in the context of post-growth.

The German education system is governed by a federalist structure, meaning that each federal state develops and follows its own specific education plans and guidelines recorded in syllabi and textbooks. As our research project was conducted in Lower Saxony and thus only involved students from secondary schools in this federal state, other states will not be considered in the subsequent executions.

Although the core curricula of Lower Saxony for geography courses in Grades 5 to 10 explicitly mention ESD as a central guideline, those grades are also tasked with establishing basic knowledge about how ESD is approached (Kultusministerium 2015a). Due to the complexity of post-growth as a topic, upper secondary schools are more suitable for dealing with this content. Based on the core curricula of the upper grades, the students become familiarised with different cultural, ecological, economic, political, and social developments and their mutual dependencies in the context of the 2030 Agenda (Kultusministerium 2017). The core curriculum of the politics and economics subject is based around and references ESD, offering opportunities for integrating post-growth within consumption decisions and their ecological, economic, social, and legal implications regarding sustainability alongside examining our economic system (Kultusministerium 2015b). In Grade 11, the students get to analyse in detail the different SDGs within the topic of global political and economic processes. Further possible topics for integrating post-growth are the opportunities of the social market economy and the chances and risks of global interdependences (Kultusministerium 2018).

Generally, the syllabi are broadly formulated but offer points of reference for post-growth, so it is also worthwhile to evaluate textbooks. Upon examining the different geography textbooks of Lower Saxony more closely, it was found that the topics of alternative economic approaches and post-growth are each included in only a few upper secondary school textbooks. For example, the Diercke Praxis book for Grade 11 in Lower Saxony includes a page on transition towns and post-growth economics with the heading "Measures for a Sustainable Spatial Development" and outlining a critical reflection on our current capitalistic economic system (Eberth et al. 2017). Another politics and economy textbook in Lower Saxony (for Grade 10) also integrates this in a reflection at the end of the section questioning the growth paradigm, titled "The Role of the State in the Social Market Economy". Via the "sustainable tetragon", the textbook page thematises the sustainability of state finances, the social sustainability of educational opportunities, and ecological and economic sustainability, discussing how they would occur by decoupling from the growth paradigm and resource consumption (Bethke et al. 2022). Another example is a politics and economy textbook in Lower Saxony for the upper grades that include sustainable ways of economising, green growth, degrowth, steps into a post-growth society and the connection between prosperity and economic growth (Becker et al. 2020). According to these findings, interdisciplinary teaching would bolster the use of these textbooks. It turns out that post-growth is not the focus within politics and economy textbooks, but a detailed evaluation whether this topic is integrated in schoolbooks and teaching materials is not possible in the course of this work. Nonetheless, due to the variety of different topics and contexts as well as the curricular guidelines, a detailed thematisation of this topic is unlikely.

Consequently, due to their general wording, the core curricula of Lower Saxony for geography and the politics and economy subject were found not to mention post-growth or other alternative approaches to our current economic growth system. The curricula in Lower Saxony, Germany, do not address the challenges of our current economic system and alternative approaches, which contradicts the emphasis placed on education's role in achieving the 17 SDGs (UNESCO 2017). As the curricula offer no concrete specifications, each school or teacher can choose whether to integrate those topics into their lessons. However, the textbooks' thematisation increases the likelihood of teachers bringing up this issue in their school lessons.

This chapter focuses on two research questions from our research project; each examines young people's suggestions for school education in the context of post-growth:

- How do the participants assess the ability of change to contribute to a "system change"?
- What ideas do the participants have for addressing post-growth in geography lessons?

Research Design and Methodology

This research project's design consisted of three online meetings held in 2020 and 2021. As this period corresponds with the COVID-19 pandemic, the meetings were implemented remotely via Webex, an online videoconferencing platform. The first and final meetings were conducted as focus groups (Bedford and Burgess 2001) (see Fig. 2), which is "a research technique that collects data through group interaction on a topic determined by the researcher" (Morgan 1996, 130). Generally, six to eight participants are invited to a focus group discussion (Krueger and Casey 2000), but this can diverge, so focus groups with four or ten participants are also feasible (Fern 1982). The size of the seven total focus groups in this study varied from three to four participants each (see Table 1). Online focus groups can be conducted more flexibly and can thus be adapted much better to the individual requirements of the participants. Online platforms offer many advantageous features for the moderator and co-moderator, such as audio and video recording of the discussion and the option to use a second screen for a second speaker (Stewart and Shamdasani 2017).

In total, 25 15- to 17-year-old upper secondary students from the Hanover region in Northern Germany participated in the survey. We chose participants in this age group because, as outlined above, the core curriculum till Grade 10 focuses on conveying basic knowledge about ESD topics (Kultusministerium 2017), and the topic post-growth only occurs, if at all, in upper grades in school books. There was no relation between the gender and the argumentation, so the gender relation of the participants is not a matter of concern (see Table 1).

When designing our study, we assumed most participants had not been in contact with post-growth-related topics, as derived from the core curriculum and textbooks, as outlined in Sect. "Status Quo of Post-growth in Secondary Schools of Lower Saxony, Germany". Therefore, the first set of focus group meetings focused on system knowledge and were structured with a prepared guideline and different stimuli. Those were, for example, a picture of a Fridays for Future demonstration; the slogan "System Change, Not Climate Change"; a video excerpt of a speech by Greta Thunberg at the COP 24 climate conference in Katowice; diagrams on the Anthropocene and *The Great Acceleration* by Steffen et al. (2015); and four short texts by (1) Tim Jackson (excerpt from *Prosperity Without Growth*), (2) Naomi Klein (excerpt from the Green

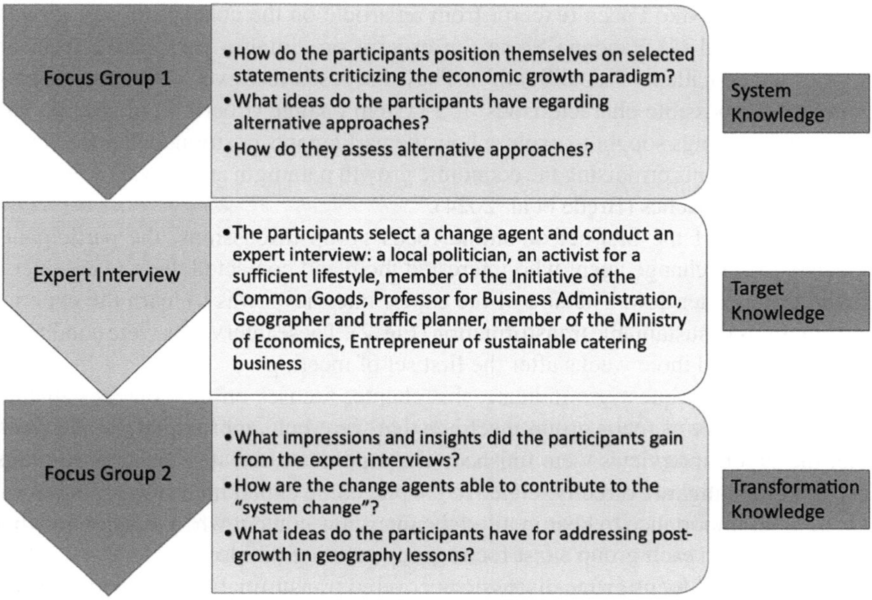

Fig. 2 Three-part structure of the survey (own illustration). Each focus group meeting was recorded and transcribed. The data were analysed using the qualitative content analysis approach by Kuckartz (2019) and MAXQDA software

Table 1 Overview of the study sample

	Number of participants	Gender distribution	Change agent for the expert interview
Group 1	4	4 male, 0 female, 0 other	A local politician
Group 2	3	0 male, 3 female, 0 other	An activist for a sustainable lifestyle
Group 3	4	2 male, 2 female, 0 other	Member of the Initiative of the Common Goods, Hanover
Group 4	3	3 male, 0 female, 0 other	Professor for Business Administration
Group 5	4	4 male, 0 female, 0 other	Geographer and traffic planner
Group 6	4	4 male, 0 female, 0 other	Member of the Ministry of Economics of Lower Saxony
Group 7	3	1 male, 2 female, 0 other	Entrepreneur of a sustainable catering business
Total	25	18 male, 7 female, 0 other	7

New Deal), (3) Niko Paech (excerpt from an article on the concept of post-growth economics), and (4) Vandana Shiva (an excerpt providing a perspective from an actor of the so-called "Global South"). They discussed the texts' different perspectives and the possible characteristics of a system change (Eberth et al. 2022). The first set of meetings sought to explore how the participants positioned themselves on selected statements criticising the economic growth paradigm and how they assessed alternative approaches (Brede et al. 2024).

At the end of the first set of online focus group discussions, the participants selected a local change agent who interested them and contacted them to conduct a digital expert interview. The aim of the expert interviews was to learn the experts' contributions to sustainable transformation (Fig. 2). These interviews were conducted between two and three weeks after the first set of meetings.

As its research questions indicate, this chapter focuses on the content gathered from the final set of focus group meetings that were held approximately one week after the expert interviews were finished. The arguments and statements during this last set of meetings are directly related to the preceding expert interviews. Therefore, it is of great importance to bear in mind the interview content when examining what was discussed in each group's last focus group meeting (Table 2).

The last set of focus group discussions covered transformation knowledge, especially regarding geography education, so these meetings were intentionally less structured than the first ones to create an open space for the participants' ideas. The first half of these meetings aimed to reflect on the expert interviews and the insights gained about post-growth. The participants were asked to explain their impressions from the interview, summarise the most critical aspects, and describe how the work of the change agents can contribute to a system change. The second half of the meetings was intended to explore the participants' ideas on how to teach post-growth geography education, thus tackling whether post-growth-related topics already take up space in school education and, if not, how incorporating post-growth into geography lessons may be accomplished.

Perspectives of Young People on Post-Growth in Geography Education: Selected Results of an Empirical Study

How Do the Participants Assess the Ability of Change Agents to Contribute to a "System Change"?

The participants deeply respected and sought out exchange and contact with the external change agents. In four groups (G2, G5, G6, G7), participants mostly referred to them as role models that others, like themselves, can learn from and imitate. Overall, it has to be said that the participants of group seven clearly attach the highest importance of all groups to change agents and role models. One of the most mentioned aspects describing change agents is their ability to contribute to a system

Table 2 Overview of the expert interviews

	The thematic focus of the expert interview
Group 1	• Sustainable living, construction, travelling, traffic (train vs. plane) • Gender equality (gender-sensitive language, gender pay gap) • Exploitation by big businesses (Nestlé) • Possibilities of action (consumerism) • Re-use and recycling (to-go cups, repair-cafés)
Group 2	• Sustainable economies and businesses • Visions for the future: fair working conditions, connection to ourselves and others, climate justice, meaningful work, co-living • Self-effectiveness: consumerism, sustainable travelling, recycling • Interaction with climate change deniers/people who lack understanding
Group 3	• Common-good orientation (sustainability, social justice and solidarity, complications with certifications) • The importance of political measures, such as subventions for sustainable businesses, implementing and regulating labels (e.g. Fairtrade) and their criteria (transport routes, production conditions…) • The importance of education/school in the context of sustainable economies for younger generations as well as older people already in charge of taking action
Group 4	• Urge for growth: Not all businesses must grow, the misbelief that prosperity equals growth, defining post-growth (focus on working conditions, salary) • Redistribution of salaries, the need for political measures • Production conditions (e.g. of smartphones), competitiveness, the lifetime of electronic products paired with the desire for economic growth of businesses, predetermined breaking points
Group 5	• Urban planning, green areas, public transportation (expenses, expansion) • Public transportation in rural areas, car dependency
Group 6	• Climate-friendly infrastructure and sustainable transportation of goods (trains, ships, cars, planes), the limited capacity of public transportation • Cheaper tickets for public transportation, the reliability of public transportation
Group 7	• The production of ice cream, transport routes for ingredients, regional ingredients • Sustainable packaging and production • Mergers with regional and small businesses

change and the motivation they create for others to become active. In this, change agents serve as essential sources of information for other people to learn how to change their own behaviour or acting. It is crucial for the participants to not only get confronted with theoretical considerations regarding a system change but also to get to know actual stories of success.

> I think that he is a high motivation for other businesses, and he shows that and how it is possible to do something that he does. Others always think that it is not possible or too difficult. But his actual implementation shows that everything is possible like that and everybody can make it in this way like him. (Canan_G7_24)

This source of information embodied through the change agents is of great importance as participants in two groups (G2, G7) complain about a knowledge deficit in this context. In the participants' opinions, this deficit must be overcome for them to become active.

> I think that knowledge transfer is the most important step. Many people simply do not know how to do their part for a sustainable lifestyle and towards a system change. And that is how people assume that they have to waive everything. (Sanja_G2_73)

Throughout the focus group discussions, it emerged that the participants value their self-efficacy. Nonetheless, there are differences between the groups regarding the mention of their self-efficacy (Fig. 3). The digits on the lines show the numbers of mentions of self-efficacy by the participants.

In this context, when participants discussed external change agents, they not only referred to prominent role models like Greta Thunberg but also to people in their immediate environment, such as friends, family members, or teachers.

> Sometimes a simple comparison is enough; sometimes you are a role model yourself. And teachers, for example, as well. When the teacher is someone who eats meat only once a

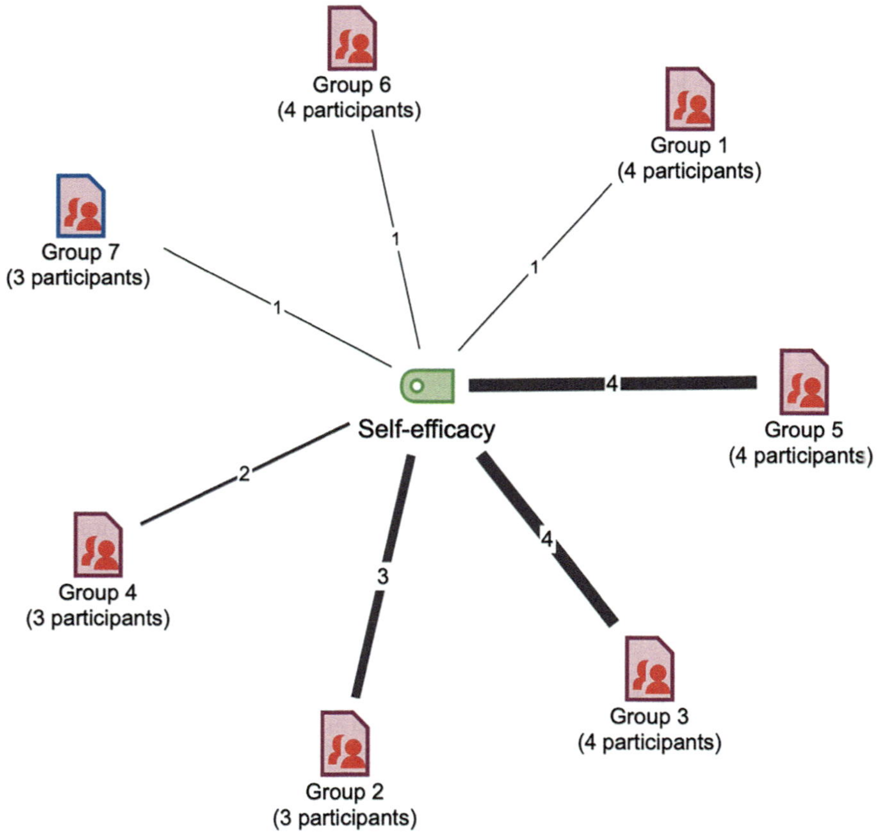

Fig. 3 Number of participants' statements in direct context of self-efficacy (own illustration)

> week, then that may already be enough for us students to get an insight into the practice. We can see how he does it and learn from it. (Timon_G6_35)

> When I notice that my friend Moritz is trying to act sustainably and is making an effort, and Rafael, for example, is trying as well, then this would rub off on me, and I will make an effort, too. It is a common project in some way. (Tarek_G5_60)

In the context of getting active themselves, the participants mentioned different fields of possibilities for action. The most mentioned aspects were waste separation (G2, G3), using bikes or public transportation instead of cars (G2, G3, G5, G6), and reducing meat consumption (G2, G3, G4).

> After hearing from the change agent, I tried to change something myself. I can say that I do not eat Nutella anymore. I also try to reduce my consumption of meat, and I do not use plastic bags anymore, in supermarkets, for example. I even get a guilty conscience when I do something like that. I simply try to do anything [...]. But everybody has to decide to what extent they want to reduce their own consumption. (Soraya_G3_57)

> Everyone can start with just little things and can contribute. For example, using the bike instead of something else when it is a short distance. (Moritz_G5_66)

In addition to mentioning individual responsibility and the possibilities of action in the context of systemic change, the participants of the three groups also mentioned the need for changes to the economic system (G2, G3, G4) and political measures (G1, G4, G7).

> Yes, many people of our generation are vegetarians and think about acting environmentally friendly. And that, of course, makes a certain impact. But nonetheless, I think that we have to consider different political actions and economic measures. I think that it is an interplay of those different things that can make a change. (Danil_G4_73)

Although the participants feel motivated and informed through contact with the change agents, the young people of two groups (G2, G7) mention their dissatisfaction with older generations and their wish to be heard and perceived. They feel smiled at and not taken seriously by older people, which overshadows their initial desire to become active. In this context, the participants often refer to Greta Thunberg and the Fridays-for-Future movement, which many people have seen as a pretext for skipping school.

> I think that it is sad that children who are interested in sustainability are being ridiculed by older people. Greta Thunberg, for example, became a meme. She was not taken seriously by many people. I think that this is sad. I think that people who are interested in sustainability topics are not only interested in themselves but more in the common good. One should not be making fun of them, but should be thankful for that. (Lia_G7_62)

What Ideas Do the Participants Have to Address Post-growth in Geography Lessons?

As mentioned, the participants criticised their deficit in knowledge of sustainable economic approaches and systemic change. In relation to this, participants in five groups (G2, G3, G4, G6, G7) complained about the school system, saying that the topics it covers are outdated and only deal with matters superficially without establishing their relation to everyday practices.

> Maybe education has to act in such a way that people are really made aware of what they can do [and] what brings change. Not only [that, but] you have to buy this and that, and you are not allowed to fly anymore. It should be discussed how to change those things and teach people to do something. Creating attention through this for this topic will help to change the economy. (Emilia_G2_72)

Participants in four groups (G1, G2, G4, G7) demanded that sustainability be discussed on different levels; for example, the nuances of post-growth economies should be discussed much more during geography lessons. Those are, for the most part, those participants who complain about the knowledge deficit in this context. Some participants mentioned that they had not had many geography lessons thus far. Regarding sustainability in school education, the participants expressed the desire to learn more about social and economic aspects and current scientific research in this field. They offer the criticism that focusing on sustainability starts in Grade 10 at the earliest, suggesting a continuous confrontation with this topic from primary school onward, as there are already small possibilities of action in younger years. Following the core curriculum of Lower Saxony, the participants state that aspects of sustainable mobility or renewable energies and consumer behaviour are covered during geography lessons. Consequently, they agree that discussing sustainable ways of economising comes off badly.

> The focus of this topic in school was often put on political measures towards renewable energies and staying competitive. In this context, we talked just a little bit about the principles of our free-market economy – for example, who has to be taxed or are subsidies yielding something – and, of course, our consumerism and that we have to think about what we use. But we never actually talk about how this is economically and sustainably feasible without losing prosperity and workplaces. That is exactly the fact that was left out. I think that this is sad. In my opinion, sustainability is a highly diverse topic, but the focus in school is always only put on one or two specific aspects. (Lars_G4_68)

In the context of wishing for a stronger focus on alternative economic approaches, they reflect on their contact with the change agents through expert interviews. As mentioned, the participants worship the contact and discussion with the change agents. Concerning school education, participants in four groups (G2, G5, G6, G7) think that local change agents should be invited to their geography lessons, for example, to show that a sustainable lifestyle does not mean waiving everything. In this context, role models' importance is frequently mentioned in relation to the participants' self-effectiveness. Furthermore, participants in five groups (G1, G3, G5, G6, G7) thought that visiting sustainable businesses would help to provide a better insight into how they economise. A visit from external change agents or a trip to a particular business is also described as a welcome change and supplement to the theoretical analysis based on texts, for example. Furthermore, the possibility of conducting a discussion with an external and involved person is seen as a beneficial competence.

> When integrating the topic of sustainable ways of economising, there should be trips to those companies to see how they work. There is no need to go abroad, but inside Germany, we could visit, for example, an ice-cream parlour that works environmentally friendly. (Canan_G7_48)

> I think that there should be more trips in the context of several topics in school anyway. But regarding sustainable economies, there should be trips to companies all the more. It would be nice to see that the school is making an effort to cooperate with sustainable businesses and establish connections and contacts for us with those businesses. That would be motivating as well. (Samira_G7_50)

In addition to expressing their desire to contact change agents and sustainable businesses, some participants mentioned that they did not understand why geography is the only subject dealing with sustainability aspects. As post-growth overlaps with politics and economics, participants of two groups (G5, G7) suggest integrating aspects in at least those two subjects. Moreover, they also mention examples connecting factors in other subjects, thus linking interdisciplinary and transdisciplinary teaching and learning principles (Table 3).

> I think it would make sense to refer to sustainability in each subject. I do not know how to do it in math… But for example, in German lessons, when we have to do a poem analysis, this could be on the topic of sustainability. (Lia_G7_44)

> We only discuss climate change or sustainability in detail in geography and a little bit in politics and economy. I have never heard of it in other subjects, and that is why I think that it can be integrated into school teaching much stronger. I mean, in other subjects apart from geography and politics. Each lesson lasts 45 minutes, and I do not really understand why this has to be separated into the different subjects instead of combining them and creating a lesson that is actually exciting and diverse. (Canan_G7_45)

Nonetheless, six participants out of four groups (G3, G4, G5, G6) are satisfied with the structure and content of their geography classes. They say that climate change and sustainability have been discussed extensively, that they feel well-informed

Table 3 Central ideas and demands upon the school in the context of post-growth (own illustration)

	Group 1	Group 2	Group 3	Group 4	Group 5	Group 6	Group 7	Total
Interdisciplinary teaching	0	0	0	0	1	0	2	3
Excursions to sustainable businesses	1	0	1	0	2	1	2	7
Stronger focus on thetopic sustainability	2	7	0	1	0	0	11	21
Motivation, showing sustainable options for action	0	2	3	2	0	6	3	16
Invite change agentsto school and classes	0	3	0	0	1	1	4	9
Total	3	12	4	3	4	8	22	56

about those topics, and that they would not neglect other geographical issues in return. However, the argumentation of the participants who were satisfied with the current geography education was primarily based on their great satisfaction with their geography teacher.

Discussion

This study found that the participants worship the contact with change agents and consider it beneficial for themselves for several reasons. If one thinks of change agents as individuals, groups, or companies functioning as role models and creating motivation for a change (WBGU 2011), the expert interviews (i.e. the interviews with change agents) had different effects on the participants. Of the expert interviews, the ones with the most mentions of the importance of change agents as role models and their importance for their self-effectiveness were those the participants conducted with an activist who supports a sustainable lifestyle, a member of the Initiative of the Common Goods, a geographer and traffic planner, and an entrepreneur who founded a sustainable catering business. Throughout the interviews with these change agents, the participants were made aware of specific possible actions in their everyday lives and got to know a story of success. This is vital to support young people's self-efficacy (Anderson 2012).

Furthermore, the participants who interviewed an activist about an appropriate lifestyle and an entrepreneur of an ice-cream business expressed the most desire for improvements at school regarding post-growth and sustainability. They sought a stronger focus on sustainability in school, more motivation, further demonstrations of possibilities for action, and invitations for change agents to visit schools. During the first focus group discussions, the role models were already established as crucial in building the self-effectiveness of the participants (Brede et al. 2024). As this is also connected to the central action fields of ESD (UNESCO 2020), the integration of change agents into school education should be increased to enable young people to become change agents themselves. Furthermore, the participants suggested inter- and transdisciplinary teaching on the topic post-growth. This corresponds with the whole-institution approach in the context of ESD (UNESCO 2020).

The dissatisfaction with older generations is an aspect frequently mentioned and given substantial weight in the first focus group discussions (Brede et al. 2024). In this context, Sanson and Burke (2020) find that intergenerational justice, which acknowledges the compounded impact of climate change on younger generations who have contributed less than their older counterparts, is of great importance (see also Vochocová and Rosenfeldová 2023). Young people's questioning of whether they were heard by older generations or being represented was also strengthened by how reporting is done on climate change and sustainability, with "little attention was given to future generations" and wherein "young people rarely spoke and, along with those yet to be born, were represented in ways that [obscure] the temporal and social inequalities that are built into climate change" (Graham and Bell 2021, 465).

According to Sanson and Burke (2020), limited possibilities for political engagement hamper young people's efforts. This finding corresponds with the participants' claim for political measures to implement another economic system. Based on these findings, schools should open—and school education should implement—transdisciplinary and interdisciplinary approaches for addressing complex topics like sustainability and post-growth. This is possible by integrating different actors in real-life contexts (Lang et al. 2012) and, therefore, people of older generations to strengthen intergenerational solidarity.

Conclusion

The results of our research show that some implications for geography education can be derived. Contact with change agents is of great importance for young people as they serve as a source of information, as role models and as motivation to get active themselves. Nonetheless, those change agents that practise self-determined work have a more significant impact on young people. The first focus group discussions have already shown some misconceptions from the participants regarding how prosperity is understood in connection with economic growth (Brede et al. 2024). In addition to the desires of young people for keener focuses on sustainability, motivation, and integrating change agents into school lessons, this knowledge gap demonstrates the critical need for education to enable young people to become change agents. Furthermore, contact with change agents can also contribute to intergenerational solidarity. In emancipatory ESD and transformative learning, the critical reflection of experiences and gained knowledge is crucial. Through contact with change agents, young people's ideas get confronted with alternative approaches. Own conceptions can be questioned through a focused reflection (Pettig 2021), which is why—besides integrating change agents into school education—a debriefing in this context is highly important as well.

References

Aitken SC (2019) Coloring outwith the lines of the map: children's geographies as contested subfield and practical global force. In: Skelton T, Aitken SC (eds) Establishing geographies of children and young people. Springer Nature Singapore, Singapore, pp 17–49

Anderson A (2012) Climate change education for mitigation and adaptation. J Educ Sustain Dev 6(2):191–206

Becker H, Benzmann S, Große Hüttmann M, Riedel H, Ringe K, Tessmar K, Tschirner M, Weinmann G (2020) Buchners Kompendium Politik. Politik und Wirtschaft für die Oberstufe. C. C. Buchner Verlag, Bamberg

Bedford T, Burgess J (2001) Using focus groups in qualitative research. In: Dwyer C, Limb M (eds) Qualitative research in geography. Longmans, pp 121–135

Bethke A, Detjen J, Henne C, Lahme S, Meske F, Quest T, Skaide F (2022) Mensch & Politik. Sekundarstufe I. 9./10. Schuljahr. Westermann Bildungsmedienverlag GmbH, Braunschweig

Brede L, Eberth A, Meyer C (2024) Post-growth from the perspectives of young people in Germany in the context of education for sustainable development. In: Leininger C, Hardouin M (eds) Curricular changes in geography education—meeting social and societal challenges of tomorrow. Springer, submitted

Eberth A, Häusler M, Meyer C, Reblin S, von Roux Y (2017) Diercke Erdkunde. Einführungsphase. Niedersachsen. Bildungshaus Schulbuchverlage, Braunschweig

Eberth A, Meyer C, Heilen L (2022) Fokusgruppen zu Postwachstumsökonomien im Kontext transformativer Bildung. In: Eberth A, Goller A, Günther J, Hanke M, Holz V, Krug A, Roncévic K, Singer-Brodowski M (eds) Bildung für nachhaltige Entwicklung—Aktuelle Impulse zu Digitalisierung, Inklusion und Klimaschutz. (Ökologie und Erziehungswissenschaft, Schriftenreihe der BNE-Kommission in der DGfE). Barbara Budrich, Opladen, pp 144–164

Eberth A, Meyer C, Heilen L (2023) "Economics for future" from different perspectives: critical reflections on SDG 8 with a special focus on economic growth and some suggestions for alternative pathways. In: De Lázaro Torres ML, De Miguel González R (eds) Sustainable development goals in Europe: a geographical approach. Springer Nature Switzerland AG, Cham, pp 153–167

Fern EF (1982) The use of focus groups for idea generation: the effects of group size, acquaintanceship and moderation on response quantity and quality. J Mark Res 19:1–13

Freeman C, Tranter P (2015) Children's geographies. Int Encyclo Soc Behav Sci 2(3):491–497

Gallo-Cruz S (2017) The insufficient imagery of top-down, bottom-up in global movements analysis. Soc Mov Stud 16(2):153–168

Göpel M (2016) The great mindshift. How a new economic paradigm and sustainability transformations go hand in hand. Springer, Berlin

Graham H, Bell S (2021) The representation of future generations in newspaper coverage of climate change: a study of the UK press. Child Soc 35(4):465–480

Haraway D (2016) Staying with the trouble: making kin in the Chthulucene. Duke University Press, New York

Harvey D (2020) Anti-capitalist chronicals. Pluto Press, London

Heilen L, Eberth A, Meyer C (2021) Young people's geographies in the times of Covid-19: system threat as a chance for system change? In: Travis C, Valenno V (eds) Narratives in the Anthropocene era. Il Sileno, Lago, pp 302–320

Jackson T (2009) Prosperity without growth. Economics for a Finite Planet. Earthscan, London

Kallis G (2018) Degrowth. Agenda Publishing, Newcastle upon Tyne

Kallis G, Kostakis V, Lange S, Muraca B, Paulson S, Schmelzer M (2018) Research on degrowth. Ann Rev Environ Resour 43:291–316

Klein N (2014) This changes everything: capitalism vs the climate. Simon & Schuster, New York

Koller H-C (2020) Problems and perspectives of a theory of transformational processes of Bildung. Educ Theory 70(5):633–651

Krueger RA, Casey MA (2000) Focus groups: A practical guide for applied research, 4th edn. Sage Publications Inc., Thousand Oaks, CA

Kuckartz U (2019) Qualitative text analysis: a systematic approach. In: Kaiser G, Presmeg N (eds) Compendium for early career researchers in mathematics education. Springer, Cham, pp 181–197

Kultusministerium N (eds) (2015a) Kerncurriculum für das Gymnasium—Schuljahrgänge 5-10. Erdkunde. Unidruck, Hannover

Kultusministerium N (eds) (2015b) Kerncurriculum für das Gymnasium—Schuljahrgänge 8-10. Politik-Wirtschaft. Unidruck, Hannover

Kultusministerium N (eds) (2017) Kerncurriculum für das Gymnasium—gymnasiale Oberstufe, die Gesamtschule—gymnasiale Oberstufe, das Abendgymnasium, das Kolleg. Erdkunde. Unidruck, Hannover

Kultursministerium (eds) (2018) Kerncurriculum für das Gymnasium—gymnasiale Oberstufe, die Gesamtschule—gymnasiale Oberstufe, das Abendgymnasium, das Kolleg. Politik-Wirtschaft. Unidruck, Hannover

Lang DJ, Wiek A, Bergmann M, Stauffacher M, Martens P, Moll P, Swilling M, Thomas CJ (2012) Transdisciplinary research in sustainability science: practice, principles, and challenges. Sustain Sci 7:25–43
Lambert D (2009) Introduction—part 1: what is living geography? In: Mitchell D (eds) Living geography. Exciting futures for teachers and students. Chris Kington Publishing, London, pp 1–7
Marston SA, Jones JP, Woodward K (2005) Human geography without scale. Trans Inst Br Geogr 30(4):416–432
Massey D (2008) World city (Reprint). Polity
Meyer C, Höbermann C (2021) From fast fashion to slow fashion—raising awareness of young people for sustainable production and consumption. In: Fargher M, Mitchell D, Till E (eds) Recontextualising geography in education. Springer Nature Switzerland, Cham, pp 167–183
Morgan DL (1996) Focus groups. Ann Rev Sociol 22:129–1523
O'Sullivan E (2012) Deep transformation: forging a planetary worldview. In: Taylor EW, Cranton P (eds) The handbook of transformative learning. Theory, research, and practice. Jossey-Bass, San Francisco, pp 162–177
Paech N (2012) Liberation from excess. The road to a post-growth economy. Oekom Verlag, München
Peck J (2009) Economic growth. In: Gregory D, Johnston R, Pratt G, Watts MJ, Whatmore S (eds) The dictionary of human geography. Wiley Blackwell, Malden, Oxford, pp 181–182
Pettig F (2021) Transformative Lernangebote kritisch-reflexiv gestalten Fachdidaktische Orientierungen einer emanzipatorischen BNE. GW-Unterricht 162(2):5–17
Potter R, Conway D, Evans R, Lloyd-Evans S (2012) Key concepts in development geography. SAGE, London
Sanson AV, Burke SEL (2020) Climate change and children: an issue of intergenerational justice. In: Nikola B, Christie DJ (eds) Children and peace. Springer International Publishing, Cham, pp 343–362
Schneidewind U (2013) Plädoyer für eine Bürgeruniversität. Duzmagazin 8(4):30–31
Schulz C, Lange B, Huelz M, Schmid B (2021) Post-growth geographies. Conceptual and thematic cornerstones of this book. In: Lange B, Huelz M, Schmid B, Schulz C (eds) Post-growth geographies. Spatial relations of diverse and alternative economies. Transcript, Bielefeld, pp 15–36
Singer-Brodowski M, Schneidewind U (2014) Transformative literacy. Gesellschaftliche Veränderungsprozesse verstehen und gestalten. In FORUM Umweltbildung im Umweltdachverband (eds) Krisen- und Transformationsszenarios: Frühkindpädagogik, Resilienz & Weltaktionsprogramm. Forum Umweltbildung, Wien, pp 131–140
Stewart DW, Shamdasani P (2017) Online focus groups. J Advert 46:48–60
UNESCO (2017) Education for sustainable development goals: learning objectives. https://www.unesco.org/en/articles/education-sustainable-development-goals-learning-objectives
UNESCO (2020) Education for sustainable development. A roadmap. UNESDOC Digital Library. https://unesdoc.unesco.org/ark:/48223/pf0000374802.locale=en
UN: United Nations (2015) Transforming our world: the 2030 agenda for sustainable development. Resolution adopted by the general assembly on 25 Sept 2015 (A/RES/70/1). https://sdgs.un.org/2030agenda
Victor P (2015) Growth. In: D'Alisa G, Demaria F, Kallis G (eds) Degrowth. In: A vocabulary for a new era. Routledge, New York, pp 109–112
Vochocová L, Rosenfeldová (2023) Are generations really divided by climate? Preference for conflict in fridays for future media coverage. Journalism Practice. https://doi.org/10.1080/17512786.2023.2260364
WBGU: German Advisory Council on Global Change (2011). World in transition. A social contract for sustainability. https://www.wbgu.de/fileadmin/user_upload/wbgu/publikationen/hauptgutachten/hg2011/pdf/wbgu_jg2011_en.pdf

WBGU: German Advisory Council on Global Change (2016) Humanity on the move: unlocking the transformative power of cities. https://www.wbgu.de/fileadmin/user_upload/wbgu/publikationen/hauptgutachten/hg2016/pdf/hg2016_en.pdf

The Importance of the University Environment in Shaping Competences Relating to Sustainable Development Among Geography Students in Poland and Czechia

Małgorzata Cichoń and Barbara Baarová

Abstract In times of global crises, education managers are expected to be more imaginative, going beyond international, standardised tests of knowledge and skills, and moving towards properly preparing citizens worldwide for a changing reality. By combining formal and informal learning environments within higher education—as part of a new learning culture—universities will play an increasingly important role, particularly in developing future competences, including those relating to sustainable development. This article presents an international study conducted between students of geography in Poland (UAM) and Czechia (OU), examining the influence of selected learning conditions in the academic environment on the level of competences relating to sustainable development. The research was based on 12 new competences proposed in 2022 in the GreenComp document (Bianchi et al. in GreenComp the European sustainability competence framework. Publications Office of the European Union, Luxembourg, 2022). Both Polish and Czech students declare that within the university environment, the level of competences relating to the values and complexity of sustainable development is predominantly influenced by study programmes, while for competences relating to the future and action, it is the non-academic environment. Statistical analyses confirm the regularities indicated above; however, strong positive correlations were noted between the competences of creative thinking and action and the way in which content is delivered in a study programme. The decisive factors for the development of these competences were problem-based learning and project-based learning, respectively.

M. Cichoń (✉)
Laboratory of Geography Didactics and Educational Research, Faculty of Geographical and Geological Sciences, Adam Mickiewicz University, Poznań, Poland
e-mail: cichon@amu.edu.pl

B. Baarová
Department of Human Geography and Regional Development, University of Ostrava, Ostrava, Czech Republic
e-mail: barbara.baarova@osu.cz

© The Author(s), under exclusive license to Springer Nature Switzerland AG 2025
D. Piróg and A. Świętek (eds.), *Future-Ready Geography*, Key Challenges in Geography, https://doi.org/10.1007/978-3-031-81268-2_8

Keywords Competences · Geographical education · Learning conditions · Sustainable development · University

Introduction

Sustainable development is a concept focused on achieving human development goals (Brundtland 1987), while maintaining the capacity of ecosystems to provide ecosystem services for future generations. Sustainable development has officially become one of the long-term goals of the European Union in accordance with Art. 3 Section 3 of the Treaty on the European Union. Taking into account current activities at the international level, the 2030 Agenda for Sustainable Development was created, which defined a model of sustainable development at the global level. It includes 17 Sustainable Development Goals (SDGs). With the intensification of global problems, it is assumed that the implementation of the concept of sustainable development should take place primarily through education. UNESCO (2014a) indicates that education is not only about teaching and research, but also about fostering cooperation within society, and changes should not only include students and teachers, but to all forms of relationships. This assumption therefore requires a different approach to education, aimed at preparing engaged, critical, but above all, environmentally conscious citizens who will strive for a sustainable lifestyle. According to UNESCO (2014b), these educational changes should focus on developing competences for sustainable development, where great importance is attached to the learning environment. The teaching and learning environment may have numerous dimensions, but in the case of activities for sustainable development, it refers primarily to the social dimension, because behaviours that are assigned specific meaning and value may take place in the school, local or global environment (Soga and Gaston 2016). There is an ongoing discussion in many forums as to whether these competences are best developed through formal, non-formal or informal learning experiences (including The Council of the European Union 2018). Research is being conducted on the impact of various educational experiences on decision-making, actions at home or in the workplace, but there are few studies on the impact of education on the level of preparedness for challenges arising from sustainable development (Lozano et al. 2015; Ploum et al. 2018; Elmassah et al. 2020). According to SDSN (2020), the academic environment has the greatest impact on the achievement of Sustainable Development Goals. Across the spectrum of universities, those educating graduates in interdisciplinary fields, such as geography, are particularly significant. These are the graduates who are expected to ultimately hold important positions in governmental and non-governmental organizations dealing with sustainable development issues. Therefore, the aim of this study is to determine which academic learning conditions develop competences for sustainable development among geography students at two universities in Central Europe. The research is based on examining the level of the 12 competences proposed by Greencomp—The European sustainability competence framework (Bianchi et al. 2022).

Literature Review

In September 2015, world leaders at the United Nations (UN) General Assembly unanimously approved the "Transforming our world: the 2030 Agenda for Sustainable Development", one of the most ambitious and significant global agreements of recent decades. At the heart of this agenda are the 17 Sustainable Development Goals (SDGs), divided into 5 areas (Planet, People, Prosperity, Peace, and Partnership), encompassing 169 goals aimed at guiding all countries towards collectively addressing our world's most pressing challenges by 2030.

Achieving sustainable development requires, primarily, universal education by equipping every individual with the appropriate knowledge, skills, attitudes, and values, i.e. competences, which, according to Wiek et al. (2011) enable effective action regarding real-world problems. In the literature, we find the term "competency", which comes from the USA and focuses on the behaviour, motivations, and other personal characteristics of an individual (Lester 2014). On the other hand, the term "competence" is of British origin and refers to the knowledge, practical skills, and understanding of the work environment, and is linked to outcomes (Winterton 2002; Brundiers et al. 2021). Competences can also be defined as all the skills necessary to undertake and perform cognitive and practical activities in a specific field of education, personal, and professional life (Niemierko 2004). Skills encompass the sphere of interaction in the teaching process, while competences refer more to the individual. These arise from the integration of a certain number of skills mastered efficiently and consciously enough to achieve the ability to freely, wisely, reflectively, and responsibly take specific actions (Okońska-Walkowicz et al. 2009) to adapt to various life situations (Perrenoud 1997; Cichoń and Piotrowska 2012). Key competences in lifelong learning were included in the Recommendation of the European Parliament and the Council of the European Union in 2006 and updated in 2018. The recommendations include 8 competences:

(1) communication in the mother tongue (understanding and creating information (since 2018)
(2) communication in foreign languages
(3) mathematical competences and basic scientific and technical competences
(4) IT competences
(5) the ability to learn
(6) social and civic competences
(7) initiative and entrepreneurship
(8) cultural awareness and expression.

Acquiring competences significantly differs from learning (understood as acquiring knowledge), because the effects of competences are specific behaviours resulting from knowledge, skills, and attitudes that can be taught but not developed (Weinert 2001). Since the Bologna reform in higher education, especially in the context of Education for Sustainable Development (ESD), various competency frameworks have been developed at different levels of education (e.g. de Haan 2006;

Rieckmann 2018; Riess et al. 2018). De Haan (2010) identified six competencies in the field of sustainable development: responsibility (values, ethics, reflection); emotional intelligence (cross-cultural understanding, empathy, solidarity, compassion); systemic orientation (interdisciplinarity); future orientation; personal involvement (self-motivation, motivating others, learning); and the ability to take action (participatory skills). Lozano et al. (2017) conducted a literature review on competences for sustainable development and identified 12 competences: systems thinking; interdisciplinary work; anticipatory thinking; justice, responsibility, and ethics; critical thinking and analysis; interpersonal relations and collaboration; empathy and change of perspective; communication and use of media; strategic action; personal involvement; assessment and evaluation; and tolerance for ambiguity and uncertainty. Similarly, the typology developed by UNESCO in the document "Education for Sustainable Development" includes similar competencies key to sustainable development: systems thinking, anticipatory competence, normative competence, strategic competence, collaboration competence, critical thinking competence, self-awareness competence and integrated problem-solving competence (UNESCO 2017). In 2022, a new document *GreenComp the European sustainability competence framework* (Bianchi et al. 2022) was published, in which 12 competences were designated, and divided into 4 groups: *Embodying sustainability values, Embracing complexity in sustainability, Envisioning sustainable futures, and Acting for sustainability.*

Striving to develop competences for sustainable development requires profound changes in the functioning of societies and economies and in the way we interact with our planet. At the same time, this transformation towards socially inclusive and environmentally sustainable economic development requires all sectors and actors to behave in a more collaborative, interconnected, systemic, and engaged manner (Sachs et al. 2019). Researchers agree that universities play an important role in assisting society in achieving sustainable development goals (Kurucz et al. 2017; Leal Filho et al. 2019). They do so through research and the education process, although, according to Solis-Espallargas et al. (2023), these elements are not always holistically interconnected. It is also worth emphasizing the "third mission" of universities (Eizaguirre et al. 2019). The social responsibility of science is currently one of the most emphasized areas of the functions of university, not only at the national level, but also globally. A declaration signed by 20 Polish universities in 2017 emphasized the role of universities as places for creating and transmitting knowledge about the reality around us, as well as the need to shape the social and civic attitudes of future elites that will contribute to community building, creativity, openness, communication, as well as social sensitivity and work culture. The Times Higher Education Impact rankings publish a ranking of universities every year, which succeed in achieving sustainable development goals (SDGs) around the world, as the goals are diverse, there are also many categories in the competition. On average, over 1400 universities from over a hundred countries participate, which also indicates the importance of SDGs attributed to these goals (The Times Higher 2023).

Universities develop skills that "are more important than ever in our rapidly changing society" (The Council of the European Union 2018, p. 23). These include problem-solving, critical thinking, collaborative skills, creativity, computational thinking, and self-regulation. Universities use various tools to combine these skills with expert knowledge and enable students to generate new ideas, theories, and products for sustainable development (Sady et al. 2019). There are also scientists who perceive competences focused on sustainable development as a determinant of an individual's effectiveness in society (Goleman 2006). However, in order to create a teaching and learning environment focused on sustainable development, students must be provided with both appropriate formal and informal education. Formal education includes study programmes, academic staff, and teaching methods, while informal education includes the university space and optional classes and activities at the university. According to Waltner et al. (2019) there is a lack of empirical research results on the factors conditioning the development of competences for sustainable development.

Research Methods

Based on the literature, a review of articles related to the research area was conducted to ensure the selection of appropriate competences and factors contributing to the academic environment. The research issues were correctly defined and operationalized and two universities in Central Europe were selected.

Research Area

The University of Ostrava (UO) is a public higher education institution located in the city of Ostrava, which is the third-largest city in Czechia. With approximately 9500 students, UO offers a diverse range of study programmes at the bachelor's, master's, and doctoral levels across various faculties, including the humanities, social sciences, natural sciences, medicine, and arts. The University of Ostrava is one of eight institutions in Czechia that prepares future geography teachers. UO has demonstrated a commitment to sustainable development, aligning its goals and activities with broader environmental, social, and governance (ESG) principles. It has also actively participated in the European Sustainable Development Week (ESDW) to showcase effective sustainable development actions and their practical applications.

The Adam Mickiewicz University in Poznań is one of ten research universities in Poland and is strongly focused on the development of scientific research and its internationalization. Currently, close to 29,000 students are enrolled in nearly 141 fields of study with over sixty specializations at the bachelor's, master's, and uniform master's degree levels. Nearly 1200 doctoral students are studying in doctoral studies in the Doctoral School of Adam Mickiewicz University. UAM and its sustainable development team conducts extensive activities, including the Academy of Sustainable Development. The Adam Mickiewicz University in Poznań was classified for the second time in the green *QS World University Rankings: Sustainability ranking*, taking 434th place globally and 4th place nationally.

Surveys

The basis for developing the questionnaire was the 12 competences proposed in the Greencomp report (Table 1).

To achieve the research goal, surveys were used by the students. The research was conducted using a standardized questionnaire. This technique was chosen due to a number of advantages, including the cost and time reduction associated with carrying out the study. In addition, the use of surveys allows for anonymity, which may positively affect the reliability of responses and the possibility of obtaining answers to problematic or embarrassing questions. The limitations of this technique include the lack of control over disturbing factors and the lack of confidence in the credibility of respondents and the credibility of their answers (Owens 2002).

To ensure validity and reliability the survey was conducted in two stages. In the first stage, a pilot study was carried out in the spring of 2023 among 20 individuals studying geography at the Adam Mickiewicz University in Poznań. The second stage of the survey took place from February 29th to March 7th, 2024 at the Adam Mickiewicz University in Poznań and the University of Ostrava. The students represented the second and third years of bachelor's degrees and the first year of master's studies (Table 2) with or without an educational and pedagogical module preparing for the teaching profession.

Due to the randomness and representative selection of the sample size, the results can be used to describe all students at public universities in Poland and the Czech Republic.

The questionnaire used in the study consisted of open, closed, semi-open, filtering, and metric questions. The following scales were used: Likert, dichotomous, nominal, and balanced. This allowed for the collection of data for the analysis and interpretation

Table 1 GreenComp areas, competences

Area	Competence
1. Embodying sustainability values	1.1 Valuing sustainability
	1.2 Supporting fairness
	1.3 Promoting nature
2. Embracing complexity in sustainability	2.1 Systems thinking
	2.2 Critical thinking
	2.3 Framing problem
3. Envisioning sustainable futures	3.1 Futures literacy
	3.2 Adaptability
	3.3 Exploratory thinking
4. Acting for sustainability	4.1 Political agency
	4.2 Collective action
	4.3 Individual initiative

Source Bianchi et al. (2022), pp 14–15

Table 2 Percentage of respondents for the two selected universities

	Direction	Number of students in a given field	Number of respondents	Percentage (%)
AMU	General geography	64	20	31
	Environmental management	99	32	32
	Educational module	38	24	63
UO	Political geography	63	23	37
	Geography teaching	78	33	42

on how the learning conditions present at the university contribute to the development of 12 competences in the field of sustainable development, the level of which was determined by students on a scale from 1 to 5. Subsequently, students selected one learning condition out of six related to the university environment (ecological solutions in the university space, content of study programmes, nature of compulsory classes, extra-curricular activities, method of content delivery, attitude of the academic teachers), that they believed influences the development of competences for sustainable development. However, students also had the option to choose a non-academic environment as the seventh factor influencing competence development. Within each selected condition, comprising three factors, the student determined to what extent (1 indicating a small extent, and 3 a large) each factor influenced the development of a given competence (Table 3).

Statistical Analysis

Data Analysis

Calculations were performed using IBM SPSS Statistics 29.0 to help calculate and analyze basic descriptive statistics for the competency indicators. To determine the relationship between quantitative variables, Pearson's r or Spearman's ρ correlation analysis was performed for ordinal variables. To determine the relationship between categorical variables, an analysis was performed using Pearson's χ^2 test or Fisher's exact test when the expected size was less than 5. To compare two groups in terms of quantitative variables, an analysis was performed using the t-test for independent samples (or its nonparametric equivalent, the Mann Whitney U-test), and when there were more groups to compare, a one-way analysis of variance. Tuckey's HSD test was used as a post-hoc test. A standard threshold of $\alpha = 0.05$ was used as the significance level.

Table 3 Learning conditions and factors used in the research procedure

Learning conditions	Learning factors
Ecological solutions in university space	1. The building infrastructure incorporates ecological solutions (light, water, and heating sensors; wall and window insulation; waste bins) 2. The surroundings include landscaped greenery (e.g. parks, gardens), with rainwater tanks, insect houses, bird houses 3. The campus and accompanying infrastructure is consolidated in one place, minimizing resource consumption (e.g. water, fuel, packaging, energy)
Content of study programmes	1. The study programme content is focused on career preparation 2. The study programme content is interdisciplinary 3. The study programme content is focused on sustainable development
Types of compulsory classes	1. Lectures and seminar meetings (consultations) 2. Classes in training rooms, laboratories, and computer labs 3. Fieldwork and visits to workplaces
Types of optional classes outside the study programme	1. Participation in popular science events (e.g. festivals, meetings with experts, job, and internship fairs, GISDay, Earth Day, Sustainable Development Week) 2. Activities for others (e.g. volunteering, charity collections, student council, trips abroad, e.g. AID) 3. Activities for science (e.g. Student Science Club, projects/grants conducted in cooperation with scientists, ERASMUS)
How to implement sustainable development content	1. Exercises using modern technology, programmes, and applications 2. Problem-based teaching (problem-solving through hypothesis formulation, discussion, simulation games) 3. Project method (creating creative solutions in a group)
Personal attitude of teachers towards sustainable development	1. The teacher provides current (but pessimistic) facts about sustainable development 2. The teacher encourages personal and responsible engagement in sustainable development 3. The teacher demonstrates their sustainable lifestyle personal attitude and example
Non-academic environment	1. Activities in the family home and neighbourhood 2. Activities in primary and secondary school, workplaces 3. Activities in the virtual world

Table 4 Descriptive statistics for competency indicators

	M	*Mdn*	*SD*	*Sk*	*Kurt*	*Min*	*Maks*
Competences 1	4.03	4.33	0.80	– 1.35	2.31	1.00	5.00
Valuing sustainability	3.78	4.00	0.93	– 0.71	0.54	1.00	5.00
Supporting fairness	4.11	4.00	0.99	– 1.13	0.92	1.00	5.00
Promoting nature	4.20	4.00	0.95	– 1.41	1.97	1.00	5.00
Competences 2	3.69	3.67	0.76	– 0.74	1.40	1.00	5.00
Systems thinking	3.82	4.00	0.96	– 0.72	0.54	1.00	5.00
Critical thinking	3.80	4.00	0.92	– 0.67	0.55	1.00	5.00
Framing problem	3.44	3.00	0.86	– 0.10	– 0.30	1.00	5.00
Competences 3	3.79	4.00	0.80	– 0.39	– 0.39	1.67	5.00
Futures literacy	3.82	4.00	1.05	– 0.48	– 0.62	1.00	5.00
Adaptability	3.83	4.00	1.05	– 0.52	– 0.75	1.00	5.00
Exploratory thinking	3.73	4.00	0.97	– 0.51	– 0.45	1.00	5.00
Competences 4	3.14	3.00	0.91	– 0.08	– 0.38	1.00	5.00
Political agency	2.93	3.00	1.20	0.16	– 0.77	1.00	5.00
Collective action	3.17	3.00	1.02	0.00	– 0.40	1.00	5.00
Individual initiative	3.31	3.00	1.21	– 0.12	– 0.95	1.00	5.00

Descriptive statistics

Table 4 presents the basic measures of descriptive statistics for the competency indicators analyzed. For the main categories of competencies (competencies 1–4), an overall index was calculated by averaging the ratings made for their three components. In the subsequent part, the indices of evaluations of the factors evaluated within each competency are presented. For all the indicators analyzed, the values of skewness were in the – 2, 2 range, indicating that the distributions for the quantitative results were close to a normal distribution. The kurtosis values did not indicate the presence of outliers. Therefore, when the other assumptions were met, the analyses were conducted based on parametric tests.

Results

The level of the 12 competences for sustainable development among Polish students ranges from medium (2.94) to high (4.36), similarly for Czech students, from 2.91 to 4.10 (Fig. 1).

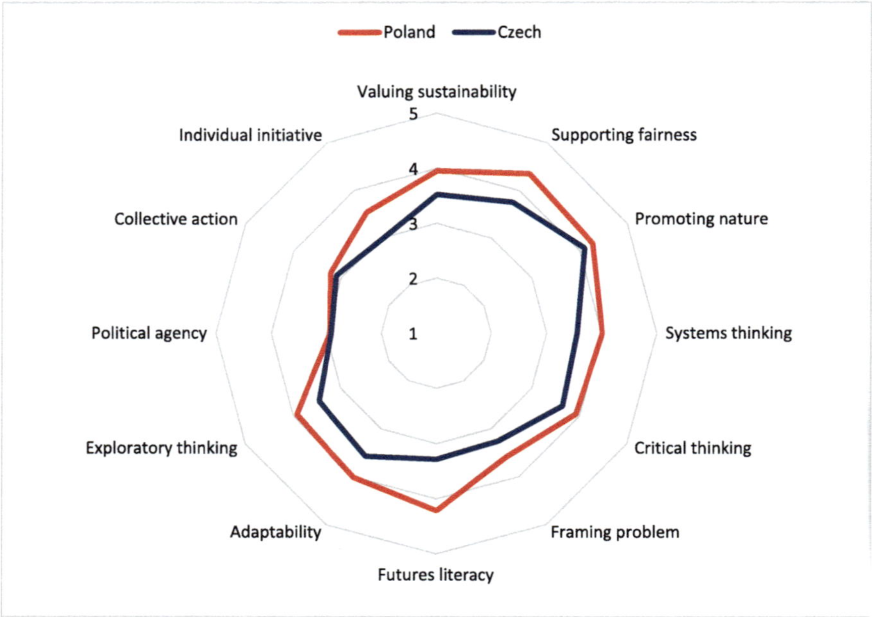

Fig. 1 Level of 12 competences in the field of sustainable development for students of the geographical faculties at the Adam Mickiewicz University in Poznań and the University of Ostrava, legend 1-low level, 5-high level

According to students from Poland and the Czech Republic, the academic and non-academic environments are of similar importance for developing specific competence groups (Fig. 2). However, considering only the learning conditions in the academic environment, differences between students can be observed (Figs. 3, 4, 5, and 6).

For the first group of competences (Embodying sustainability values), Czech students assign three times more importance to the content in study programmes, while Polish students emphasize both study programmes and the methods of implementing the content during compulsory classes (Fig. 3).

For the second group of competences (embracing complexity in sustainability), which include systems thinking, critical thinking and problem framing, both Polish and Czech students attach significant importance to study programmes (approximately 30% of respondents). Every fifth Polish student believes that attendance at compulsory classes plays a significant role in developing competences from the second group, while every sixth believes that the methods of implementing the content are important. In the case of Czech students, the percentage of individuals declaring the importance of each of the six learning conditions is similar (Fig. 4).

The opinions of students regarding the impact of individual learning conditions on the level of competences related to envisioning sustainable futures (competences from group III) are similar (Fig. 5). However, it is worth emphasizing that for both

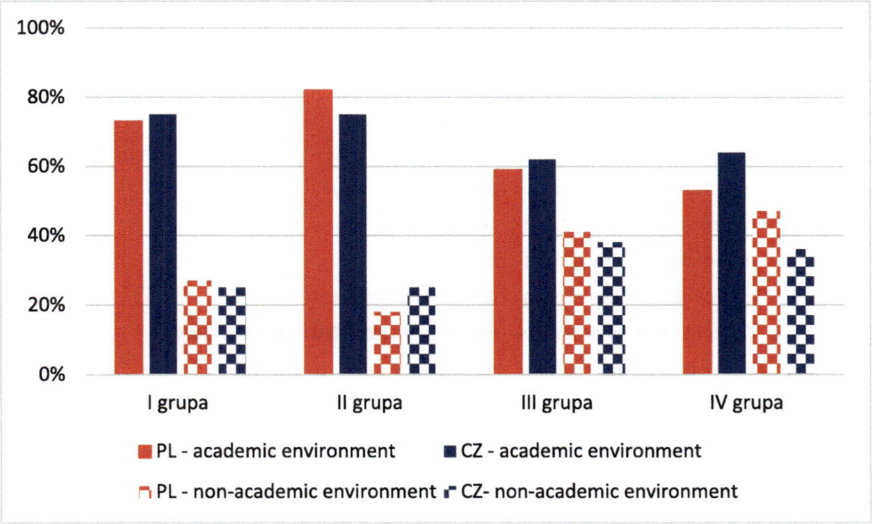

Fig. 2 Importance of the academic and non-academic environment in developing competences for sustainable development according to surveyed students

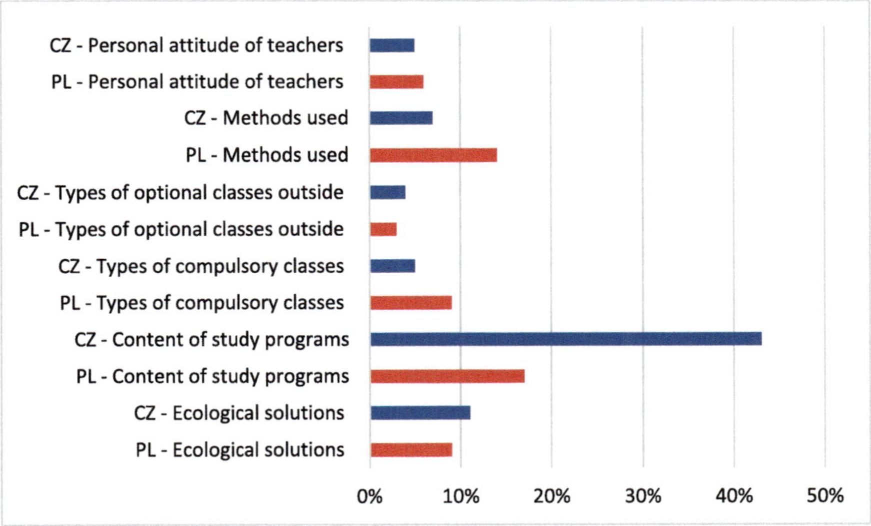

Fig. 3 Percentage of students declaring the importance of 6 selected learning conditions for developing group I competences related to SD values

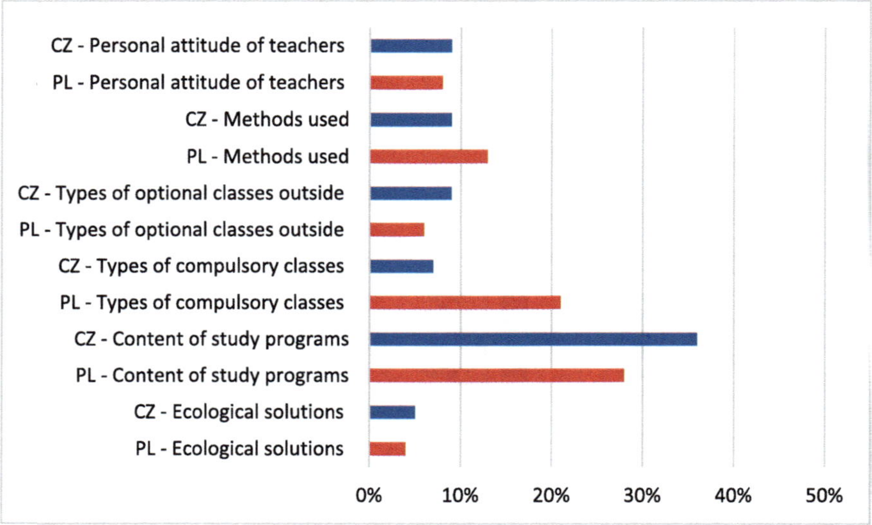

Fig. 4 Percentage of students declaring the importance of 6 selected learning conditions for developing group II competences related to the complexity of SD

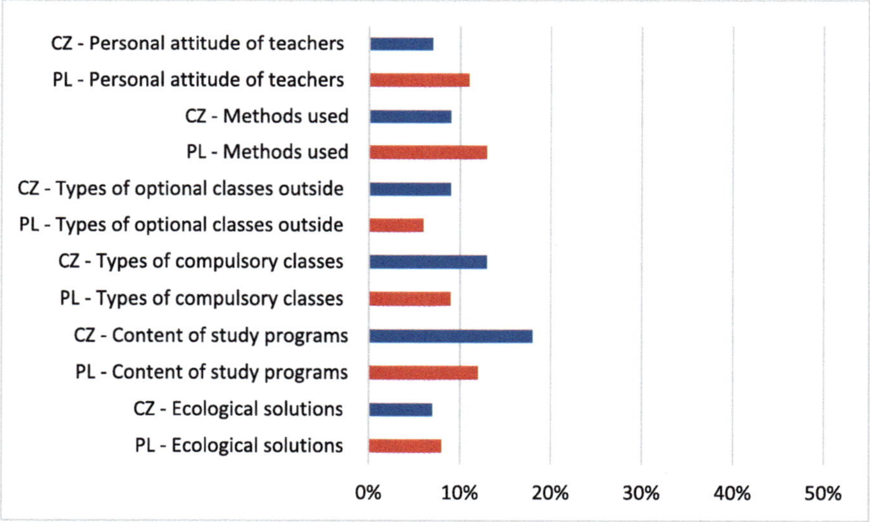

Fig. 5 Percentage of students declaring the importance of 6 selected learning conditions for developing group III competences related to the future of SD

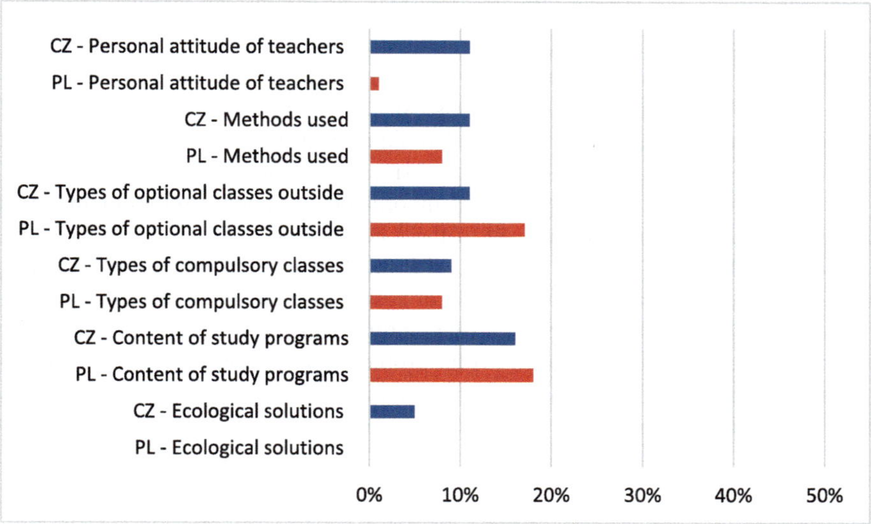

Fig. 6 Percentage of students declaring the importance of 6 selected learning conditions for developing group IV competences related to SD activity

Polish and Czech students, the non-academic environment may also have a significant impact (Fig. 2).

For the fourth group of competences, Polish students highlighted the following as being of decisive importance: Acting for sustainability, extra-curricular activities related to the organization of departmental and university-wide events, and the content in study programmes. In turn, the percentage of Czech students declaring the importance of individual learning conditions is similar. It is worth noting the significant difference with the importance of the teacher's attitude—with 11% for Czech students and only 1% for Polish students (Fig. 6).

A Spearman correlation analysis between the assessment of 6 learning conditions and the level of 12 competences related to sustainable development revealed several strong and positive correlations (Table 5). The competence of valuing sustainability was strongly and positively associated with the assessment of the methods of content delivery. A higher rating of the content in study programmes (moderate positive correlation) and the non-academic environment (moderate positive correlation) was linked to a higher level of supporting fairness. A higher rating of the non-academic environment was associated with higher levels of the promoting nature competence (moderate and positive correlation). A higher rating for the study programme was associated with higher levels of problem framing (moderate and positive correlation). In the third group of competences, positive correlations ranging from moderate to strong were observed between the non-academic environment and adaptability, as well as between the assessment of content delivery methods in study programmes in the field of sustainable development and exploratory thinking. Higher ratings for the specific learning conditions were associated with higher levels of competence. In the fourth group of competences, positive and moderate correlations were observed between the assessment of the non-academic environment and all competences: political agency, collective action, and individual initiative. Strong and positive correlations were also observed between the assessment of content delivery methods and political agency and collective action. No significant associations were observed between the remaining factors and the assessment of competence levels.

It was noted that the higher the importance assigned to the methods of content delivery, the higher the level of the 4 competences: Valuing sustainability, Exploratory thinking, Political agency and Collective action (Table 5). Considering the content delivery methods in the field of sustainable development, competences from group three (Futures literacy, Adaptability, and Exploratory thinking) were positively and strongly correlated with "problem-based learning", while competences from group IV, related to acting for sustainability, were strongly and positively correlated with "project-based learning" (Table 6).

Discussion

The growing importance of competences for sustainable development constitutes a demanding area of action for higher education institutions, as the academic environment is perceived not only as a place of education, research, and innovation, but also as a venue for social learning (Ferrer-Balas et al. 2010; Salvioni et al. 2017). The offer of university education should respond to the requirements of rapidly changing societies, as universities should prepare not only professionals with specific disciplinary skills, but above all conscious citizens guided by the principles of sustainable development.

According to the results obtained, the majority of students indicated that significant importance for developing SD competences lies in appropriate study programmes oriented towards SD content, which was also confirmed by moderate correlations. Similar results were obtained by Hammer and Lewis (2023), where both students, graduates, and internship supervisors concluded that study programmes should not only focus on specialized and methodological competences, but also on communication and action for sustainable development. Research results obtained by others, Cebrián and Junyent (2014), Lozano et al. (2019), and Solis-Espallargas et al. (2023), encourage us to take these competences into account. This is indeed a valid statement considering the results presented in this study, where the level of students' competences in the field of action is lower than the level of other competences. In addition to action, there is significant importance in the academic environment for developing the competences of reflection and creative thinking. Therefore, it seems understandable that there is a consensus that competences should constitute the overarching framework in the development of study programmes in the future (Weinert 2001; Mochizuki and Fadeeva 2010; Lozano et al. 2019; Wilhelm et al. 2019; Hammer and Lewis 2023).

Table 5 Spearman correlations between the assessment of factors and the level of 12 competences regarding sustainable development

		Competences 1			Competences 2			Competences 3			Competences 4		
		Valuing sustainability	Supporting fairness	Promoting nature	Systems thinking	Critical thinking	Framing problem	Futures literacy	Adaptability	Exploratory thinking	Political agency	Collective action	Individual initiative
Eco-infrastructure	*r*	0.27	− 0.26	0.07	− 0.58	− 0.69	0.02	0.14	0.10	0.41	–	0.00	− 0.50
	Mr	0.419	0.433	0.844	0.225	0.130	0.977	0.700	0.775	0.236	–	1.000	0.667
Study programmes	*r*	0.29	0.49	0.21	0.11	0.20	0.34	0.22	0.14	0.46	0.11	− 0.09	0.12
	Mr	0.053	< 0.001	0.156	0.508	0.213	0.027	0.375	0.579	0.050	0.614	0.670	0.587
Compulsory activities	*r*	− 0.03	0.16	0.33	− 0.14	0.34	0.21	0.24	− 0.31	− 0.46	− 0.08	− 0.36	− 0.36
	Mr	0.930	0.582	0.248	0.556	0.139	0.368	0.408	0.286	0.101	0.821	0.280	0.275
Optional activities	*r*	− 0.26	0.63	0.89	− 0.03	0.13	0.51	0.62	− 0.32	0.17	− 0.01	0.07	0.08
	Mr	0.742	0.368	0.106	0.941	0.728	0.131	0.057	0.375	0.636	0.966	0.790	0.746
Non-academic environment	*r*	0.7	0.38	0.44	0.04	0.20	0.13	0.27	0.44	0.26	0.36	0.47	0.42
	Mr	0.706	0.024	0.008	0.828	0.299	0.522	0.056	0.001	0.065	0.007	< 0.001	0.001
Ways of implementation	*r*	0.54	− 0.06	0.36	0.32	0.16	− 0.09	0.04	0.34	0.51	0.79	0.74	0.43
	Mr	0.039	0.822	0.188	0.250	0.569	0.751	0.882	0.213	0.050	0.001	0.004	0.139
The attitude of academic teacher	*r*	0.29	0.53	0.51	0.31	0.15	− 0.01	0.03	0.48	− 0.08	0.71	0.49	0.40
	Mr	0.486	0.176	0.196	0.354	0.665	0.994	0.931	0.114	0.803	0.072	0.264	0.371

Table 6 Spearman correlations between competence level and factor scores

	Competences 1		Competences 2		Competences 3		Competences 4	
	r_s	p	r_s	p	r_s	p	r_s	p
The content in the study programme is focused on work preparation	0.28	0.060	0.19	0.224	0.33	0.174	0.05	0.823
The content in the study programme is considered interdisciplinary	0.15	0.340	0.19	0.225	0.42	0.073	0.03	0.908
The content of the study programme is focused on sustainable development	**0.43**	**0.003**	**0.21**	**0.188**	**0.11**	**0.656**	**– 0.01**	**0.976**
Exercises using modern technology, programmes, and applications	0.11	0.688	0.15	0.605	0.12	0.665	0.38	0.198
Problem-based learning (problem-solving through hypotheses and discussion, simulation games)	**0.39**	**0.147**	**0.01**	**0.985**	**0.64**	**0.009**	**0.47**	**0.103**
Project method (group development of creative solutions)	**0.21**	**0.456**	**0.07**	**0.814**	**0.27**	**0.326**	**0.85**	**< 0.001**

Statistically significant correlations (p < 0.05) are given in bold

In addition to acquiring knowledge, skills, and attitudes in the field of SD, students should be able to solve real problems in a critical, reflective way (Wals 2011; Wiek et al. 2011). Moreover, according to Glasser and Hirsh (2016), the definition for competences already implies that the combination of knowledge, skills, and habits should aim at posing problems and solving them. Additionally, a university geography graduate is more often expected to be mathematically and technologically competent to succeed in the job market (Cichoń and Piotrowska 2018). According to Delamare Le Deist and Winterton (2005), the descriptions of competences for university graduates in study programmes are often abstract and lacking context, while, challenges and tasks related to sustainable development often gain significance in a specific work environment (Ploum et al. 2018). This means that different competences will be required from teachers than from officials. Therefore, the context of the work environment is an important factor, but sustainability competences should take precedence over workplace requirements. Hammer and Lewis (2023) emphasize that study programmes should be based on a transdisciplinary approach, introducing students to new forms of thinking and action. According to Leal Filho et al. (2018), universities could introduce appropriate policies and practices such as green campuses, SD training, and joint local activities. However, in reality, it is easier to support SD by implementing interdisciplinary learning and methods such as inquiry-based learning, projects, and problem-solving methods. Research conducted among students at the University of Ostrava (Václavík et al. 2022) focused on the quality of higher education, focusing mainly on strategies, and learning outcomes. This led to a deeper examination of learning attributes, including teaching strategies, methods, and forms, to determine their level of support by students. The findings suggest that students prefer active approaches to learning, especially heuristic and activation methods, for their ability to promote critical thinking and shape thought processes. A very important element they underlined was the use of digital technology. A significant difference was noted between the methods preferred by students and those implemented in practice, highlighting the need to align educational approaches more closely with student preferences. Sady et al. (2019) emphasize that universities have the right conditions in this respect, because by combining skills with expert knowledge, they enable students to generate new ideas, new theories, and new products for sustainable development.

The above conclusions are particularly important in the context of the obtained research results, where the increase in competences among students of geography is strongly correlated with the methods of delivering content in study programmes. Competences from group three (related to future literacy, adaptation, and creativity) were positively and strongly correlated with problem-based learning, while competences from group four, associated with action, were strongly and positively correlated with project-based methods.

Project-based learning is a dynamic classroom approach where students actively explore real-world problems and challenges, acquiring a deeper knowledge through active exploration of complex questions, problems, or challenges. Renowned educators and scholars like John Dewey and Jean Piaget have laid the foundations for this approach, emphasizing the importance of learning through doing. This methodology encourages students to engage in projects that are both meaningful to them and reflective of real-world issues, facilitating not only academic learning but also critical life skills such as collaboration, problem-solving, and time management. Specifically, Bolstad et al. (2023) examined students' self-assessment of the integration of aspects of sustainability in their projects and their development of the normative, critical thinking, and self-awareness competences from the UNESCO key competences for sustainability depending on the project topic.

Similarly, problem-based learning emphasizes student-centred inquiry and problem-solving strategies to address complex problems. Howard S. Barrows, for instance, championed the idea that problem-based learning could significantly enhance the critical thinking and problem-solving skills of students by engaging them in the process of finding solutions to open-ended problems. This method fosters an environment where students learn to collaborate, research, and apply their knowledge to develop practical solutions, preparing them for the unpredictability and complexity of real-life situations. The 2030 Agenda for Education for Sustainable Development (ESD) emphasizes the transformation of educational practices to address sustainability issues through interactive, student-centred pedagogies such as problem-based learning (PBL), particularly in South Africa (Sousa and Golightly 2023), where the geography curriculum and national guidelines encourage developing critical thinking, problem-solving skills, and a commitment to sustainability among students. They illustrate how the integration of ESD with PBL in geography education promotes transformative social learning by addressing complex interdisciplinary real-world problems.

Both methodologies, while distinct in their execution, share the common goal of preparing students for the challenges of the modern world by promoting a deep, active learning experience. As education continues to evolve, the integration of project-based and problem-based learning represents a forward-thinking approach to developing the skills and competencies necessary for success in the twenty-first century (Cichoń and Piotrowska 2012, Piotrowska et al. 2019). In this context, the dilemma of how to shape competences for sustainable development is becoming increasingly important. There are many strategies that can be applied to the context of contemporary environmental problems (Piotrowska et al. 2022). However, many researchers, including Sainz (2021), emphasize that the issue of sustainable development must serve methodological and pedagogical solutions and vice versa.

The main difficulty faced by educators of sustainable development results from interdisciplinarity (Eizaguirre et al. 2019), and the incorporation of global issues and technologies into curriculum requirements (Tarozzi and Inguaggiato 2016). Furthermore, it is important that students have frequent contact and socialization through interactions with the environment and develop an attitude of respect towards nature as a cultural and social norm. These assumptions regarding the implementation of the concept of sustainable development can be found in geographic education programmes. Geographic education, is aimed at enabling learners to: "*(a) learn about the world created by nature and culture, including science, art and technology; (b) prepare to change the world by developing physical and mental skills, abilities, interests and passions, as well as self-educational needs and skills; c) shape an individual personality by developing creative attitudes and a personal attitude to moral, social, cognitive, artistic and religious values*" (Okoń 2001, p. 191). The aim of geographic education, implemented both formally and informally, is actions aimed primarily at the development of cognitive (knowledge), practical (skills) and sensory-emotional (attitude) activities, taking into account geographical content that describes and allows an understanding of the surrounding world. Geographic education aims to achieve a comprehensive understanding of how the environment functions at the local, regional, or global scale in relation to humans. We look at the environment from the perspective of global problems, which Wilczyńska-Wołoszyn (2014) calls the determinants of the geographic education system. The most important determinants include social and cultural changes, technological progress, and global environmental problems. Geographic education, due to its interdisciplinary nature and relationship with the environment, will play a key role in implementing the concept of sustainable development. Geography graduates will therefore be well-prepared in the field of sustainable development. This is evidenced by the results obtained on the level of competences, which are mostly high-level competences, especially in the areas of values and future thinking. Therefore, the opinion of E. Szkurłat and other authors that the core educational value of geography results from integrating content about the natural environment with socio-economic and humanistic issues, is completely understandable (Szkurłat et al. 2017). In this context it is necessary to mention that a revision of the educational programme framework in the Czech Republic has been ongoing for several years. One public consultation took place in 2022, and a second public consultation is currently underway in April and May 2024. From the school year of 2027/2028 onwards, the implementation of new school management programmes (RVP) will begin. Geography is being proposed as one of the ten educational areas of basic education (NPI 2024). Each area has its own educational branches, and in the case of geography, it is one of the four branches of Sustainability of Life on Earth. Graduates of geography education at the University of Ostrava are expected to teach this branch much more intensively than has been done so far. This is why it is important to target the development of SD competencies in the university environment and develop competences so that they are not removed from reality.

Conclusion

In the context of the global challenges facing the modern world, it has become clear that education for sustainable development is not an option, but a necessity. To meet these challenges, it is imperative that universities engage in an ongoing process of thoughtful educational reform that considers both an individual student's needs and the imperatives of global sustainability. Achieving the 17 Sustainable Development Goals requires a shift in thinking and action at all levels of society. Universities, as institutions educating the future generations of leaders and decision-makers, must play a key role in this transformation, because, as the research results show, the level of many competences is influenced by specific learning conditions in the academic environment. Universities cannot limit themselves only to scientific research but must also engage in an effective educational process through problem-based learning (PBL) and project-based learning (PBL) strategies. The academic community is expected to integrate knowledge from various fields, which is the foundation of effective education for sustainable development. Universities should promote and support collaboration between different departments and fields of study to create richer, more interdisciplinary teaching programmes.

Geographic education will play a particularly important role in these transformations due to its interdisciplinary nature, global approach to content, and the use of geo-information technology. Graduates of geographic disciplines will become leaders in education for sustainable development. In times of global environmental crises, managers of geographic studies are expected to be more courageous in properly preparing students for the new, changing reality. Academic staff responsible for teaching in geographic disciplines should be aware that the methods of delivering course content and the attitude of the academic teacher have a real impact on the level of competences developed among students. Therefore, it is worth considering introducing mandatory methodological training for employees and providing access to their scientific stations for classes/workshops for students (Piotrowska and Cichoń 2016). Research conducted among geographic students in Poland and the Czech Republic may also contribute to a university/faculty debate on the importance of the academic community in the process of developing competences for sustainable development.

References

Bianchi G, Pisiotis U, Cabrera Giraldez M (2022) In: Punie Y, Bacigalupo M (eds) GreenComp the European sustainability competence framework.. Publications Office of the European Union 2022, Luxembourg

Bolstad T, Lundheim L, Orlandić M, Strømberg A, Zimmermann P (2023) Do students reflect on sustainability? Student development of competencies for sustainability in project-based learning. Nordic J STEM Educ 7:73–83. https://doi.org/10.5324/njsteme.v7i1.4478

Brundiers K, Barth M, Cebrián G, Cohen M, Diaz L, Doucette-Remington S et al (2021) Key competencies in sustainability in higher education-toward an agreed-upon reference framework. Sustain Sci 16:13–29. https://doi.org/10.1007/s11625-020-00838-2

Brundtland GH (1987) Our common future, report of the world commission on environment and development. Oxford University Press, Oxford, p 37

Cebrián G, Junyent M (2014) Competencias profesionales en Educación para la Sostenibilidad: un estudio exploratorio de la visión de futuros maestros. Enseñanza Cienc Rev Investig Exp Didácticas 32:29–49

Cichoń M, Piotrowska I (2012) Kształtowanie kompetencji kluczowych wśród studentów geografii poprzez metodę projektu, esej geograficzny i recenzję. In: Podgórski Z, Szkurłat E (eds) Wybrane problemy akademickiej i szkolnej edukacji geograficznej. Toruń, pp 151–168

Cichoń M, Piotrowska I (2018) Level of academic and didactic competencies among students as a measure to evaluate geographical education and preparation of students for the demands of the modern labour market. Quaestiones Geographicae 37(1):81–94. https://doi.org/10.2478/quageo-2018-0006

De Haan G (2006) The BLK '21' programme in Germany: a 'Gestaltungskompetenz'-based model for education for sustainable development. Environ Educ Res 12(1):19–32. https://doi.org/10.1080/13504620500526362

De Haan G (2010) The development of ESD-related competencies in supportive institutional frameworks. Int Rev Educ 56:315–328

Delamare Le Deist F, Winterton J (2005) What is competence? Hum Resour Dev Int 8(1):27–46

Eizaguirre A, García-Feijoo M, Laka JP (2019) Defining Sustainability core competencies in business and management studies based on multinational stakeholders' perceptions. Sustainability 11:2303. https://doi.org/10.3390/su11082303

Elmassah S, Biltagy M, Gamal D (2020) Engendering sustainable development competencies in higher education: the case of Egypt. J Clean Prod 266:121959

Ferrer-Balas D, Lozano R, Hisingh D, Buckland H, Ysern P, Zilahy G (2010) Going beyond the rhetoric: system-wide changes in universities for sustainable societies. J Clean Prod 18:607–610

Glasser H, Hirsh J (2016) Toward the development of robust learning for sustainability core competencies. Sustain J Rec 9:121–134
Goleman D (2006) Emotional intelligence. Random House Publishing Group, New York
Hammer T, Lewis AL (2023) Which competencies should be fostered in education for sustainable development at higher education institutions? Findings from the evaluation of the study programs at the University of Bern, Switzerland. Discov Sustain 4(1):19. https://doi.org/10.1007/s43621-023-00134-w
https://www.timeshighereducation.com/impactrankings#!/length/25/sort_by/name/sort_order/asc
Kurucz EC, Colbert BA, Lüdeke-Freund F, Upward A, Willard B (2017) Relational leadership for strategic sustainability: practices and capabilities to advance the design and assessment of sustainable business models. J Clean Prod 140:189–204
Leal Filho W, Skanavis C, Kounani A, Brandli LL, Shiel C, do Paco A, Pace P et al (2019) The role of planning in implementing sustainable development in a higher education context. J Clean Prod 235:678–687
Lester S (2014) Professional competence standards and frameworks in the United Kingdom. Assess Eval High Educ 39(1):38–52
Lozano R, Ceulemans K, Scarff SC (2015) Teaching organisational change management for sustainability: designing and delivering a course at the university of Leeds to better prepare future sustainability change agents. J Clean Prod 106:205–215
Lozano R, Merrill MY, Sammalisto K, Ceulemans K, Lozano FJ (2017) Connecting competences and pedagogical approaches for sustainable development in higher education: a literature review and framework proposal. Sustainability 9:1889. https://doi.org/10.3390/su9101889
Lozano R, Barreiro-Gen M, Lozano FJ, Sammalisto K (2019) Teaching sustainability in European higher education institutions: assessing the connections between competences and pedagogical approaches. Sustainability 11:1602. https://doi.org/10.3390/su11061602
Mochizuki Y, Fadeeva Z (2010) Competences for sustainable development and sustainability: significance and challenges for ESD. Int J Sustain High Educ 11:391–403
Niemierko B (2004) Cele kształcenia. In: Kruszewski K (ed) Sztuka nauczania. Czynności nauczyciela. Warszawa
NPI (2024) Revize rámcových vzdělávacích programů. https://prohlednout.rvp.cz/zakladni-vzdelavani/vzdelavaci-oblasti/geo/geo/udrzitelnost-zivota-na-zemi. Accessed 1 Apr 2024
Okoń W (2001) Nowy słownik pedagogiczny. Wydawnictwo Akademickie "Żak", Warszawa
Okońska-Walkowicz A, Plebańska M, Szaleniec H (2009) O kompetencjach kluczowych, e-learningu i metodzie projektów. WSiP, Warszawa
Owens LK (2002) Introduction to survey research design. https://www.researchgate.net/publication/253282490_INTRODUCTION_TO_SURVEY_RESEARCH_DESIGN. Accessed 12 Nov 2018
Perrenoud P (1997) Construire des compétences dès l'école. ESF, Paris
Piotrowska I, Cichoń M (2016) Znaczenie stacji terenowych w kształtowaniu kompetencji naukowego badania środowiska geograficznego na różnych poziomach edukacji. In: Kostrzewski A (ed) Stacje naukowe Uniwersytetu im. Adama Mickiewicza w Poznaniu. Poznań, pp 27–34
Piotrowska I, Cichoń M, Abramowicz D, Sypniewski J (2019) Challenges in geography education—et review of research problems. Quaestiones Geographicae 38(1):61–74
Piotrowska I, Cichoń M, Sypniewski J, Abramowicz D (2022) Application of inquiry-based science education, anticipatory learning strategy, and project-based learning strategies in the context of environmental problems of the contemporary world. In: Vega Alfonso Garcia de la (ed) Didactic Strategies and Resources for Innovative Geography Teaching. IGI Global, pp 23-50, https://doi.org/10.4018/978-1-7998-9598-5.ch002
Ploum L, Blok V, Lans T, Omta O (2018) Toward a validated competence framework for sustainable entrepreneurship. Organ Environ 31(2):113–132
Rieckmann M (2018) Learning to transform the world: Key competencies. In: Leicht A, Heiss J, Byun WJ (eds) Issues and trends in education for sustainable development. UNESCO,

Paris, pp 39–59. https://www.academia.edu/35941134/Chapter_2_Learning_to_transform_the_world_key_competencies_in_ESD
Riess W, Mischo C, Waltner E-M (2018) Ziele einer Bildung für nachhaltige Entwicklung in Schule und Hochschule: Auf dem Weg zu empirisch überprüfbaren Kompetenzen. GAIA 27:298–305. https://doi.org/10.14512/gaia.27.3.10
Sachs JD, Schmidt-Traub G, Kroll C, Lafortune G, Fuller G (2019) Sustainable development report 2019. Bertelsmann Stiftung and Sustainable Development Solutions Network, New York
Sady M, Żak A, Rzepka K (2019) The role of universities in sustainability-oriented competencies development: insights from an empirical study on Polish universities. Admin Sci 9. https://doi.org/10.3390/admsci9030062
Sáinz TA (2021) Educación para el desarrollo sostenible: una visión crítica desde la Pedagogía. Revista Complutense de educación 32(2):249–259. https://doi.org/10.5209/rced.68338
Salvioni DM, Franzoni S, Cassano R (2017) Sustainability in the higher education system: an opportunity to improve quality and image. Sustainability 9:914
SDSN (2020) Accelerating education for the SDGs in universities: a guide for universities, colleges, and tertiary and higher education institutions. Sustainable Development Solutions Network (SDSN) 2020, New York
Soga M, Gaston KJ (2016) Extinction of the experience: the loss of human-nature interactions. Front Ecol Environ 14:94–101
Solís-Espallargas C, Ruiz Morales J, Valderrama-Hernández R, Alcántara-Rubio L (2023) Systematic review of sustainable development goals in higher education institutions: first five years of agenda 2030. Pedagogika 152(4):79–110
Sousa L, Golightly A (2023) Geography education for sustainable development through problem-based learning. https://doi.org/10.1007/978-981-99-2687-9_10
Szkurłat E, Hibszer A, Piotrowska I, Rachwał T (2017) Podstawa programowa przedmiotu geografia; II etap edukacyjny: klasy IV–VIII. In: Podstawa programowa kształcenia ogólnego z komentarzem; szkoła podstawowa geografia. ORE 2017, Warszawa, p 11–24
Tarozzi M, Inguaggiato C (2016) Global citizenship education in Europe, a comparative study of education policies across 10 EU Countries 2016. Research deliverable issued within the European project Global Schools. Provincia Autononoma di Trento, Trento, Italy
The Council of the European Union. Council recommendation of 22 May 2018 on key competences for lifelong learning. 2018/C 189/01. https://eur-lex.europa.eu/legal-content/EN/TXT/PDF/?uri=CELEX:32018H0604(01)&rid=7. Accessed 22 May 2018
The Times Higher Education (2023) Impact rankings
UNESCO (2014a) UNESCO roadmap for implementing the global action programme on education for sustainable development. UNESCO, Paris, FR
UNESCO (2014b) Sustainable development begins with education: how education can contribute to the proposed post-2015 goals. UNESCO, Paris, France. https://unesdoc.unesco.org/ark:/48223/pf0000230508. Accessed 10 Jan 2014
UNESCO (2017) Education for sustainable development goals: learning objectives. https://unesdoc.unesco.org/ark:/48223/pf0000247444. Accessed 11 Jan 2017
Wals AEJ (2011) Learning our way to sustainability. J Educ Sustain Dev 5:177–186
Waltner E-M, Rieß W, Mischo C (2019) Development and validation of an instrument for measuring student sustainability competencies. Sustainability 11:1717
Weinert FE (2001) Concept of competence: a conceptual clarification. In: Rychen DS, Salganik LH (eds) Defining and selecting key competencies. Hogrefe & Huber Publishers, Seattle, WA, pp 45–65
Wiek A, Withycombe L, Redman CL (2011) Key competencies in sustainability: a reference framework for academic program development. Sustain Sci 6(2):203–218
Wilczyńska-Wołoszyn M (2014) Edukacja geograficzna w australijskim systemie kształcenia. In: Szkurłat E, Głowacz A (eds) Edukacja geograficzna na świecie i w Polsce—wybrane problemy. Łódź, pp 59–72

Wilhelm S, Förster R, Zimmermann AB (2019) Implementing competence orientation: towards constructively aligned education for sustainable development in university-level teaching-and-learning. Sustainability 11:1891. https://doi.org/10.3390/su11071891

Winterton J (2002) Entrepreneurship: towards a competence framework for developing SME managers. In: United States Association for small business and entrepreneurship conference proceedings; January 2002; Reno, Nevada 2002

Defining and Measuring the Level of Sustainability for Developing World, A Review in Urban Prospective

Asad Aziz

Abstract Urban areas around the world are the hub of population and built-up infrastructure with varying environmental, political, social and spatial characteristics. The fragmentation in all these due to rapid unplanned urbanization puts significant pressure on natural resources. This creates serious issues for the ecological capacity of cities, particularly in Global South. This unplanned urban expansion challenges the urban sustainability, and past studies proved this phenomenon. This study makes an attempt to review the past literature and ground investigation of the issues regarding the urban sustainability, population, developing countries, and urbanization. After deep examination of the past literature from world reputed international journals, government documents World Bank reports, and books the study extract the results. This comprehensive review found that the unplanned urbanization, rapid population growth, spatial inequality of urban land use, urban land use fragmentation are challenging tasks to ensure the city infrastructure, development and ensure sustainability for provision of services to locals. The results of the study can be used for city planning and development for future and forceful tools for infrastructure development, urban land use and spatial homogeneity.

Keywords Development · Population · Urban · Urbanization · Sustainability

Sustainability and the Urban World

Sustainability is a broader term in multiple domains, mostly studies consider the economic, social, spatial, and environmental parameters for sustainability measurements (Pastor et al. 2023; Purvis et al. 2019; Mensah 2019). It has been investigated that cities around the world, face multi-domain problems affecting the regional sustainability and ecological pattern of cities (Hosseini et al. 2019). The concept and strategies of the sustainable city and its concept vary from time to time and space to space (Wang 2022). However, this idea gained more and more importance

A. Aziz (✉)
Department of Geography, University of Gujrat, Punjab, Gujrat P/O 50700, Pakistan
e-mail: asadazizpakistan@gmail.com

© The Author(s), under exclusive license to Springer Nature Switzerland AG 2025
D. Piróg and A. Świętek (eds.), *Future-Ready Geography*, Key Challenges in Geography, https://doi.org/10.1007/978-3-031-81268-2_9

among thinkers, leaders, and managers, for making the planet the home of man. The fundamental goal for a sustainable city is the provision of services to its citizens at maximum livelihoods and equity (Malla and Bandh 2023). A sustainable city is a different concept in developed and developing states around the world and every country is trying to attain the picture of a sustainable city even in developed nations (Hu and Li 2012). Sustainability means the minimum resource with maximum efficiency, improved quality of life without damaging the environment, social and economic stability, and equity (Wang 2012).

For environmental sustainability, the previous studies consider the indicators from the physical environment including land use land cover change, green extant in an area, air quality, and land surface temperature (Arshad et al. 2020; Lai et al. 2020). For economic sustainability measurements, the researchers use land values in an area, employment ratio in a region, size of a house, house ownership, and job opportunities in an area (Alabdulwahab 2023). Similarly, for measuring social sustainability several studies use the sense of belonging, level of maintenance of neighborhoods, neighborhood services, health care facilities, presence of parks in a settlement, sports facilities, and solid waste management in an area (Staniškienė and Stankevičiūtė 2018; Dijkstra-Silva et al. 2022). However, there still is a debate about measuring sustainability using fixed indicators, due to the complicated morphology and heterogeneous nature of different environmental indicators.

In developing world, all of the major cities are expending in every phase of social, economic, political, and ecological circles (Paulin 2019b). This development may relate to sustainable land planning, infrastructure development, capital structure, good governance, technology and innovation in transportation, management of mass transit, information technology, etc. Meanwhile in case of developing world or Global South, the situation is different due to unplanned urbanization and population growth. This is challenging for city infrastructure in term of land, municipal services, ecosystem services, environmental means and economy. In the world, artificial intelligence has been adopted by several countries even in the developed nations to face and tackle these issue and find their problems (Demaidi 2023), particularly in infrastructure and transportation, education, industries, food and agriculture, and healthcare centers, but the developing world is still lack this capacity (Paulin 2019a; Batty 2019).

Geographical Scales and Sustainability

In the world, it has been investigated that the cities are getting more and more developed in every phase of social, economic, political, ecological, and technological race (Paulin 2019b). A city is called a sustainable city, when it has ability to accept new challenges that emerge daily in a given population and maximizes its capacity to meet the issues and improve the livability of its citizens (Lin et al. 2019; Bibri and Krogstie 2019). This concept of livability can be understood by multiple themes in environmental, economic, and social domains with big data and big data analyses

(Setioningtyas et al. 2022). The fundamental element for a sustainable city is data availability and monitoring to take quick action on any emergency (Muhamad Khair et al. 2020). This emergency may relate to a natural disaster or man-made. Different strategies can be associated with data availability for example early warning before an incident, mapping the areas with high risk of any disaster, and the hotspot location of any crime or disaster with temporal variation as environmental indicators (UK 2022). Similarly, to provide the emergency response to affected community sustainable streets, one stop solutions and security at public measures with spatial data, is also an element for safe city planning (Cyclomedia 2023). Similarly, in social and economic fields a sustainable city has maximum potential to provide all social and economic benefits with maximum resilience capacity to cape the disaster.

Every country has its geographical characteristics that vary in scale and time (Mueller 2020). Therefore, it is difficult for cities of developed and developing countries to take different initiatives for sustainable movement (Torrens 2022; Lu et al. 2022; Zhang et al. 2022). Usually, sustainability is more concerned with environment, its consequences, and environmental values. These environmental values are associated with physical attributes of an area, economic conditions in a region, or social concerns within a society. All of these parameters have different indicators for the measurement of sustainability. Around the globe in developing nations, the population increases which boosts the scale and nature of urbanization from micro to macro level, and creates multiple economic, social, and environmental challenges (Siciliano 2012). This micro-to-macro level urbanization is caused by unplanned and rapid urban growth and migration from rural to urban areas in developing nations (Kleemann et al. 2017). Sustainability associated with a city called urban substantiality, is a fundamental goal to achieve UN-17 goals to ensure the quality of life (Chai and Seto 2019). Therefore, it requires attention in all spheres of urban life, including environmental, economic, and social concerns.

Sustainability and Its Pillars

Since the emergence of three basic pillars of sustainability (environment, economy, and society), social sustainability is still least defined in an urban world (Larimian and Sadeghi 2019). Quite often, social sustainability is associated with the level of quality of life and satisfaction, while "add-on" with environmental sustainability is the standard of function with natural resources and environmental services (Deleryd and Fundin 2020; Wang et al. 2022). However, measuring social sustainability and its formulation is still a challenging task for researchers, due to its multifaceted and cross-disciplinary nature in philosophical, practical, and political factors (Arundel and Ronald 2017). The preference of every geographical region or area also creates the diverse nature of this concept, its formulation, conceptualization, and practical implementation. The basic traits of a sustainable city is only limited to economic opportunities, low crime rate, sustainable land use, settlement, and residential planning, along with reasonable investment in infrastructure and development are also its subsections of sustainability pillars (Amjad and Asaad 2018).

In a community, human activities are the results of the different operations performed by community through economic, environmental, political, and social relations (Angelidou 2014). However, defining sustainability in a city or a sustainable city is commonly used in literature and it is ubiquitous (Janik et al. 2020).

For a sustainable city, the environmental, economic, and social factors in a collective sense should be incorporated and designed to create a healthy ecosystem and environment for present and future generations (Pasimeni et al. 2019; Crane et al. 2021). Along with this aspect of creating sustainability in the city, other social and ecological justice also take part in city sustainability optimizing, food, water, energy, and other consumption, to mitigate greenhouse gases by reducing carbon emission, reducing air pollution, and protecting the natural resource (Cui et al. 2021).

Sustainability and Sustainable Development

Sustainability is the fundamental principle for sustainable development. The long-term sustainable development goals need sustainability in every field and sphere of life. Sustainability is fundamental for living long and better, empowering life, minimizing environmental degradation with maximum benefits from natural resources. Some of these concerns are lowering the emission of CO_2, water pollution, and air pollution to sustain the quality of environment for sustainable living (Fu 2020). Sustainability is the ability to develop and exist for a long time with a better life quality without depleting the natural resources, that's why it is a key for better and secure future (Roy 2021). However, the population and urbanization are increasing gradually, which create diverse challenges for every country including developed and developing states around the world (Ari and Yikmaz 2019). Therefore, the scale of these challenges for the Global South is more influential due to the number of ecological, economic, social, and political reasons and instabilities. As the world population is increasing, therefore, to preserve natural resources is critically important (Zhou et al. 2018).

For sustainable development for the region and world, it is necessary to tackle ecological, social, and economic concerns affecting the sustainability for short- and long-term with multiple scales (Javanmardi et al. 2023). It is thinking of the day, for a better tomorrow in the near future (Moallemi et al. 2020). Population increase is significantly associated with the increase of built-up area and more pressure on natural resources. Along with man-made and natural such as climate change has long-term consequences on the regional stability, weather pattern, natural disasters causing flash floods, heavy rain, degradation of local ecosystem and agriculture. It is estimated that these consequences will be more frequent and severe in the future due to global climate change and changes in local weather patterns (Shah 2023).

Cities and United Nations Sustainable Development Goals

Cities are the areas for global living, according to a survey in 2023 approximately 4.4 billion people are living in cities around the world (Fernandes and Machado 2022). This trend is going on increasing gradually and it is expected that about 70 percent of the population will live in cities by 2050 (Fallah Shayan et al. 2022). On the other hand, the slum areas in urban fringes are also the living portion on Earth for the people and 1.1 billion people are living in slums which is also expected to increase by 2 billion in the next 30 years (Soma et al. 2022; Bikis 2023). The rapid rate of population increases and urbanization is far over the capacity of the urban areas to meet their demands and sustain their living condition for a healthy life and ecosystem. This rapid rate of urbanization out-space the housing infrastructure, road networking, and segregated residential colonies, which leads to the slums or slums-like pattern of an urban area or a city. Due to this happening in cities the public spaces, urban green, cityscape, and local services get limited and air pollution, poor hygiene, solid waste material, and other related issues are persisting in mage cities in the world (Wang et al. 2020). This will lead to challenges for city sustainability and also for the United National Sustainable Development Goals, particularly SDG-11 to make the cities inclusive, safe, resilient, and sustainable for human settlement (United Nation 2015a). Therefore, it is very necessary to manage, monitor, and control unplanned urbanization for urban sustainable development.

In developing nations, most of the population moved toward cities in small towns and small cities which led to inequalities and poverty for an urban area to provide resources and social equity (Xu et al. 2022). This will impact urbanization a problem for the poor and bless for the rich (Kuddus et al. 2020). Energy crises, solid waste, no access to public transportation, and insufficient public infrastructure are some of the major challenges in these areas. It is estimated that a total of 60 to 80% of energy is consumed by cities and cities contribute approximately 75% of carbon emissions around the world (Crane et al. 2023; Meng 2019). Due to this impact, the urban areas even the major cities are vulnerable to climate change and natural disasters which lead to economic, social, and human loss. On the planet Earth, urban areas cover only 3 percent of the land area but contribute to increasing natural disasters (He et al. 2021), climate change, floods, and pollution along with other social and ecological disruptions. These phenomena will lead to inequalities, social and climate gentrification, and locational unrest and insecurity for every individual at local, regional, and national to international scales (González-Marín et al. 2023).

Sustainability and Future

Urban sustainability is demanding and need for a better and sustainable future (Roggema 2016). The world getting intelligence, and efficiency in every sphere of life which demands a high level of living in every corner of a region (Lu et al.

2023). Therefore, a healthy environment, pure air without any pollution, and land with all ecosystem services are necessary for a sustainable city (Bănică et al. 2020). Creating every city, a sustainable city is the agenda of every developed nation to ensure the quality of their citizens by providing adequate housing, safe and clear food and water, provision of ecosystem services, city green, accessible transportation, affordable energy, industrial development without or, low carbon emission and environmental justice (Ibiwumi and Evangelisca 2022). This scale of development will ensure the standard of living and surely make the city a sustainable city. This city will have more resilience to cope with any natural disaster with low loss of economic, societal, or ecological infrastructures (Abdul and Yu 2020).

Sustainability is diversity from geographical to natural aspects (Olanrewaju et al. 2018). Geographical aspects include its approaches from local to regional and international, while natural aspects include its morphology from rural to urban sustainability (Qureshi et al. 2023). Similarly, its background is varied from biological to ecological and social to economic at given scales. Sometimes, due to diversity in societal background, the planners also include the political sustainability for regional peace and justice for security across the continents (Chung et al. 2020). Therefore, every aspect and every dimension of sustainability play a fundamental role in development and defining the characteristics of an area for the future (Rossi 2021).

Biological and ecological sustainability, environmental, and ecosystem play a vital role in environmental sustainability (Lin et al. 2021). It will improve the quality of the ecosystem, ecosystem services, and health of the environment, ensure our health, and protect the natural resources for future generations. However, going green and sustainable is challenging in the current era of unplanned development, extensive usage of natural resources, industrialization, and rapid transportation (Dastjerdi and Nasrabadi 2022; Olanrewaju et al. 2015). The overuse of oil and gas leads to more emission of carbon and carbon dioxide into the atmosphere which will lead to an increase in atmospheric temperature and global warming.

The questions associated with sustainability like what we are, where we go, and what is next are important questions for our lives and future generations along with other species on the planet Earth and biodiversity (Myers 2019). Our actions, decisions, and works have a long-term impacts on our future generations to define their lives on Earth planet (Daya-Winterbottom 2023). In this regard practicing sustainability means we ensure the ethical choices and think for the future that brings a healthy and safe future for everyone and that has a positive impact on every individual in a society. A planned transportation system, social justice, environmental sustainable development, and sustainable business practices lead to education on the loss of natural resources, saving energy, and providing nature base solutions for urban development (Wang et al. 2021). All of these sustainable actions will create a sustainable society and quality lifestyle for the people, which make a real difference in the ecosystem. If these approaches are adopted by the region, definitely provide long-term impacts to the region and even the entire world.

Urban Expansion and Sustainability

It has been predicted that by 2050 the world population may reach 10 billion so there needs to be more space to live, more food to live, and more resources for their survival (Kumar et al. 2022). This increasing number of urban consumers need more resources which ultimately upscale the level of dependency on the environment. This overuse of natural resources causes the depletion of natural resources, loss of biodiversity, environmental degradation, and climate change. All of these issues are challenges for sustainability and sustainable development and need targeted action (United Nation 2023).

Urban expansion and sustainability are associated (Soliman and Soliman 2022). Sustainable urban development is a fundamental pillar for achieving the UN goals (Mustafa et al. 2021). Sustainable urban development means meeting the demands of urban dwellers without compromising future generations to meet their demands (Yussif et al. 2023). Urbanization or urban expansion is a global phenomenon that shapes and transfers the world into new spatial domains with both positive and negative impacts and shapes the area with varying geographical characteristics (Iyer 2020; Zhang and Su 2016). To turn urbanization into sustainable urbanization, it is critical to ensure the urban life quality and tackle the environmental challenges (Wei et al. 2023). Sustainable urbanization is focused on building cities with economically vibrant, and also have a balance between the environmental, ecological, political, social, and economic needs of urban dwellers (Adu-Gyamfi et al. 2021). These cities have a strong network of urban planning and resource management that is responsive to the needs of their residents, government, stakeholders, and urban managers. Sustainable cities can be created by investing in sustainable ideas to reduce the dependency on natural sources and empower the economic scale of a country (Najihah et al. 2021). These cities create jobs for people, reduce environmental degradation, and ensure the quality of life. Additionally, these cities improve the living conditions of local areas and create opportunities for social inclusion among various sectors of society.

The growing population demands more space for infrastructure development and makes the city more congested and populated (Broere 2016). When the cities invest more in innovative ideas, these cities become more resilient, greener, and clear for a longer time (Zhou et al. 2022). These cities have significant impacts on the environment by reducing pollution, waste production, and energy consumption (Addas 2023b). Sustainable smart cities have efficient transportation networks with urban green spaces and their associated infrastructure, which decrease air pollution, and improve air quality and carbon emissions (Shahid et al. 2021). This will ultimately reduce the negative impacts of climate and climate change (Addas 2023a).

It is a matter of fact that much of the population around the world, including developing and developed countries moving toward cities to improve quality of life and economic opportunities (Din et al. 2023). The urban areas on the other hand have limited capacity to cope with the demands of the locals at a time (Musa et al. 2023). These demands are associated with environmental, economic, ecological, political, or social needs. Environmental demands include the provision of fresh air, water, and green infrastructure, the ecological demands include shelter for living, oxygen for living, land for agriculture, and a sustainable environment for better life quality. The economic demands include job opportunities and social demands include the sense of belonging, social justice, and security. However, the growing demands of the population are challenging for cities as well as for humans (Anthony 2023).

Building a sustainable city is challenging, because of the heterogeneous nature of urban areas, which ultimately create varying scales of urban agglomerations (Bobylev and Sterling 2016). This heterogeneous nature does not have any specific nature of sustainability indicators even in the past studies, therefore it is also difficult to measure their level of sustainability (Dronova 2017). Anyhow, most nations around the world are committed to tackling the problems and solving them by 2030 in promoting sustainable development visions and achieving the United Nations' sustainable development goals. This includes the provision of equal access to basic services and eradicating life-threatening poverty, particularly in the Global South and developing regions. Rapid urbanization creates different threats to global prosperity, human economic development, and regional security and its consequences are uneven (Gennari and Navarro 2019). This will lead to low socio-economic productivity, poor neighborhoods, severe inequalities (Zhou et al. 2021; Brelsford et al. 2017).

Measuring Sustainability

Measuring sustainability is an important question and need of the day. Measuring the sustainability for sustainable development. This measurement can be done by selecting different indicators. These indicators indicate and track the progress level, evaluate the efficiency of actions taken to make sustainable development, and communicate the goals for the future. Measuring sustainability is not only a formal duty, it is a challenging task for every stakeholder, researcher, scientist, urban manager, and decision-maker. This is challenging due to its diverse background and multiple dimensions including ecological, environmental, and social spheres in an ecosystem. Every city or geographical area has its different issues and challenges with different resilience and risks (Shamsuddin 2020).

Measuring sustainability is a set of data collection, processing, investigation, and analyses to know how sustainability is or something more. As the morphology of every region is different that's why it is difficult to quantify (Munda 2015). Sometimes it is impossible to measure for some regions. Several methods and strategies are available for measuring sustainability including personal observations and lab experiments. However, it depends on the nature of observations data availability and proposed objectives. In environmental sustainability, most methods quantify the ecosystem services, environmental values, and environmental conditions of the local area for example daily weather conditions, air quality, emission of CO_2, and extant biodiversity (Husgafvel et al. 2015). In the case of economic sustainability, the analysis involved the job opportunities in an area, land and housing rent, transportation, and industrial conditions in an area. While in the case of social sustainability, it is more concentrated on promoting justice, equity, sense of belonging, social security, and high quality of life (Doğu and Aras 2019). In developing countries, in the case of social and economic domains mostly personal observations and surveys are used for data collection or analyses of socio-economic profile in the region or regions. Environmental sustainability is measured by using computer-aided techniques and instruments.

Sustainable Development and United Nations Agenda 2023

To gain in-depth knowledge of sustainability and its implementation, it is very necessary to study and research the United Nations sustainable development goals and agenda 2023. The UN Agenda 2023 is a successor of the Millennium of Goals which is divided into 17 goals. The UN adopted these goals in 2015 (United Nation 2015b), which are as follows:

1. **Ending poverty**: The covid-19 impacted this goal. It is estimated that in 2020, 119 to 124 million fell into extreme poverty due to its impacts worldwide.
2. **Ending hunger**: According to this goal 70–161 million people are currently suffering from hunger around the globe.
3. **Ensuring good health and fostering the well-being of all age groups**: This goal is proposed to ensure the quality of life and well-being of age groups worldwide with equal facilities for health and health emergencies.
4. **Ensuring inclusive and equitable quality education**: This goal is associated with quality education for all across the globe without any political, social, or economic discrimination.
5. **Achieving gender equality and empowering all women and children**: Poor families will be treated as equal for all, ensuring the quality of life, gender equality, and children's health.
6. **Sustainable water and sanitation for all**: This goal ensures pure water and sanitation conditions for all.
7. **Affordable energy**: To reduce carbon emissions and ensure the quality of air for a sustainable environment and ecosystem, affordable energy and its provision are fundamental for this technological and industrial era. Adopt this goal for reducing the usage of fossil fuels and carbon emissions in the atmosphere.
8. **Sustainable economic growth**: is necessary for regional and global sustainable development. Developed and developing states are cooperating to achieve this goal.
9. Sustainability in infrastructure: in a need for urban areas around the world to cover and maintain the population pressure to reduce social inequality.

10. **Reducing inequality**: among the people and the nation is required for developing the world for sustainable development.
11. **Making the cities more resilient**: According to United Nations data approx. 156 nations around the world, developed national policies for making the urban areas more resilient and stable.
12. **Sustainable production and consumption**: Worldwide multiple policies are implemented across the borders for sustainability in production and consumption for economic gain and political stability.
13. **Urgent actions to tackle climate change**: Nations around the world are trying to achieve the zero level of carbon emissions to combat climate change and global changes in weather patterns.
14. **Life under seawater**: The biodiversity and life under the sea and ocean are fundamental goals for sustainability for the conservation of natural resources and human life on planet Earth.
15. **Life on land**: The planet Earth has been home to man for billions of years. Changing patterns of geographical approaches in every corner of the world create different issues and challenges for human beings. Ensuring the quality of life on land is the need of the day for well-being and prosperity.
16. **Peace, justice and strengthen the institutes**: Cooperation among the intuitions and nations for important integration regional development, and economic gain for property and economic strength.
17. **Partnership for the goals**: The UN goals are the need for every individual from local to regional and national to international levels. The partnership is needed to gain the goals and sustainability for regional peace, justice, economic profit, and making the Earth the home of men.

Conclusion

Sustainable development in urban areas around the world getting more and more important to increase resilience. Scientists, urban managers, planners, developers, ecologists, and economists around the world, strive to reduce the carbon and ecological footprints to protect the environment and maintain the real morphology of the urban ecosystem. Sustainability is fundamental for the economic, environmental, ecological, political, and social resilience in the cities for the provision of services, and equity among the communities for all individuals. It will establish a healthy relationship between biotic and abiotic components for a better future. Sustainable urban development ensures economic prosperity, social security and equity, economic growth, health, and justice for all citizens of an area. This can only be possible by adopting environment-friendly strategic approaches to reduce the negative impacts of human society and its dependency on nature and the ecosystem. Reducing carbon footprints, introduction of solar planners and sole carbs are the best examples of such smart technologies which are adopted by different nations.

Sustainable practices have diverse backgrounds and classifications in every field and area, the most effective and environmental concern for sustainable urbanization is the concept of green roofs and in-built infrastructure. This green infrastructure helps to reduce the temperature by regulating the air quality and maintaining the temperature in a dense urban network. For economic concerns, the concept of circular economy regulates the economic cycle from a single individual to a cycle model. Similarly, the concept of reuse, reducing, and regrading the material improves human efficiency to meet their demands and reduce the burden on natural resources. Measuring sustainability is challenging, due its diversity and heterogeneous morphology of indictors and parameters, every geographical scale has its own environmental, social, ecological and political priorities. Therefore, it is necessary to tackle these issues on top priorities base and solve to ensure the quality of life. For every country, the UN Sustainable development goals should be the part of planning and developing in making the city for the home of man, this is only the solution to make the planet Earth the home of men and achieving the UN goals in true spirit.

References

Abdul L, Yu T-f (2020) Resilient urbanization: a systematic review on urban discourse in Pakistan. Urban Sci 4(4)

Addas A (2023a) The concept of smart cities: a sustainability aspect for future urban development based on different cities [Original Research]. Front Environ Sci 11. https://doi.org/10.3389/fenvs.2023.1241593

Addas A (2023b) The importance of urban green spaces in the development of smart cities [Original Research]. Front Environ Sci 11. https://doi.org/10.3389/fenvs.2023.1206372

Adu-Gyamfi M, He Z, Nyame G, Boahen S, Frempong MF (2021) Effects of internal CSR activities on social performance: the employee perspective. Sustainability 13(11):6235

Alabdulwahab S (2023) The impact of a sustainable economic development focus on the real exchange rate in Saudi Arabia. Sustainability 15(18)

Amjad A, Asaad A (2018) Introductory chapter: overview of sustainable cities, theory and practices. In: Amjad A, Asaad A (eds) Sustainable cities. IntechOpen, pp Ch. 1. https://doi.org/10.5772/intechopen.82632

Angelidou M (2014) Smart city policies: a spatial approach. Cities 41:S3–S11

Anthony B (2023) The role of community engagement in urban innovation towards the co-creation of smart sustainable cities. J Knowl Econ. https://doi.org/10.1007/s13132-023-01176-1

Ari I, Yikmaz RF (2019) Chapter 4—Greening of industry in a resource- and environment-constrained world. In: Acar S, Yeldan E (eds) Handbook of green economics. Academic Press, pp 53–68. https://doi.org/10.1016/B978-0-12-816635-2.00004-3

Arshad A, Ashraf M, Sundari RS, Qamar H, Wajid M, Hasan M-u (2020) Vulnerability assessment of urban expansion and modelling green spaces to build heat waves risk resiliency in Karachi. Int J Disast Risk Reduct 46:101468. https://doi.org/10.1016/j.ijdrr.2019.101468

Arundel R, Ronald R (2017) The role of urban form in sustainability of community: the case of Amsterdam. Environ Plann b: Urban Anal City Sci 44(1):33–53

Bănică A, Kourtit K, Nijkamp P (2020) Natural disasters as a development opportunity: a spatial economic resilience interpretation. Rev Region Res 40(2):223–249. https://doi.org/10.1007/s10037-020-00141-8

Batty M (2019) Foreword: the landscape of smart cities. In: Mora L, Deakin M (eds) Untangling smart cities. Elsevier, pp ix–xi. https://doi.org/10.1016/B978-0-12-815477-9.09987-4

Bibri SE, Krogstie J (2019) Generating a vision for smart sustainable cities of the future: a scholarly backcasting approach. Eur J Futures Res 7(1):5. https://doi.org/10.1186/s40309-019-0157-0
Bikis AA-O (2023) Urban air pollution and greenness in relation to public health. J Environ Public Health (1687-9813 (Electronic))
Bobylev N, Sterling R (2016) Urban underground space: a growing imperative: perspectives and current research in planning and design for underground space use. Tunnell Undergr Space Technol 55:1–4. https://doi.org/10.1016/j.tust.2016.02.022
Brelsford C, Lobo J, Hand J, Bettencourt LMA (2017) Heterogeneity and scale of sustainable development in cities. Proc Natl Acad Sci 114(34):8963–8968. https://doi.org/10.1073/pnas.1606033114
Broere W (2016) Urban underground space: solving the problems of today's cities. Tunnell Undergr Space Technol 55:245–248. https://doi.org/10.1016/j.tust.2015.11.012
Chai B, Seto KC (2019) Conceptualizing and characterizing micro-urbanization: A new perspective applied to Africa. Landsc Urban Plann 190:103595. https://doi.org/10.1016/j.landurbplan.2019.103595
Chung S-H, Kim S-J, Park S-Y, Kim JH (2020) Past, present, and future of social housing in Seoul: where is social housing heading to? Sustainability 12(19)
Crane M, Lloyd S, Haines A, Ding D, Hutchinson E, Belesova K, Davies M, Osrin D, Zimmermann N, Capon A, Wilkinson P, Turcu C (2021) Transforming cities for sustainability: a health perspective. Environ Int 147:106366. https://doi.org/10.1016/j.envint.2020.106366
Crane M, Lloyd S, Haines A, Ding D, Hutchinson E, Belesova K, Davies M, Osrin D, Zimmermann N, Capon A, Wilkinson P, Turcu C (2023). Transforming cities for sustainability: a health perspective. National Library of Medicine (1873-6750 (Electronic))
Cui W, Wan A, Xin F, Li Q (2021) How does carbon emission reduction efficiency affect regional income inequality? The mediator effect of interregional labor flow. Math Probl Eng 2021:5578027. https://doi.org/10.1155/2021/5578027
Cyclomedia (2023) World-leading cities use smart visual data to improve safety. Your guide to safe cities. https://www.cyclomedia.com/en/safe-city-landing-page
Dastjerdi HK, Nasrabadi NH (2022) Interrelationships between urban policy and climate, with emphasis on the environment. City Territ Arch 9(1):27. https://doi.org/10.1186/s40410-022-00165-2
Daya-Winterbottom T (2023) Defining our legacy to all future generations. J R Soc New Zealand:1–14. https://doi.org/10.1080/03036758.2023.2241375
Deleryd M, Fundin A (2020) Towards societal satisfaction in a fifth generation of quality—the sustainability model. Total Qual Manage Bus Excellence:1–17. https://doi.org/10.1080/14783363.2020.1864214
Demaidi MN (2023) Artificial intelligence national strategy in a developing country. AI and Soc. https://doi.org/10.1007/s00146-023-01779-x
Dijkstra-Silva S, Schaltegger S, Beske-Janssen P (2022) Understanding positive contributions to sustainability. A systematic review. J Environ Manage 320:115802. https://doi.org/10.1016/j.jenvman.2022.115802
Din AU, Ur Rahman I, Vega-Muñoz A, Elahi E, Salazar-Sepúlveda G, Contreras-Barraza N, Alhrahsheh RR (2023) How sustainable transportation can utilize climate change technologies to mitigate climate change. Sustainability 15(12)
Doğu FU, Aras L (2019) Measuring social sustainability with the developed MCSA model: Güzelyurt case. Sustainability 11(9)
Dronova I (2017) Environmental heterogeneity as a bridge between ecosystem service and visual quality objectives in management, planning and design. Landsc Urban Plann 163:90–106. https://doi.org/10.1016/j.landurbplan.2017.03.005
Fallah Shayan N, Mohabbati-Kalejahi N, Alavi S, Zahed MA (2022) Sustainable development goals (SDGs) as a framework for corporate social responsibility (CSR). Sustainability 14(3)
Fernandes D, Machado C (2022) Chapter 9—Connecting ecological economics, green management, sustainable development, and circular economy: Corporate social responsibility as the

synthetic vector. In: Machado C, Davim JP (eds) Green production engineering and management. Woodhead Publishing, pp 183–236. https://doi.org/10.1016/B978-0-12-821238-7.00001-4
Fu B (2020) Promoting geography for sustainability. Geogr Sustain 1(1):1–7. https://doi.org/10.1016/j.geosus.2020.02.003
Gennari P, Navarro DK (2019) The challenge of measuring agricultural sustainability in all its dimensions. J Sustain Res 1(2):e190013, Article e190013. https://doi.org/10.20900/jsr20190013
González-Marín P, Vásquez-Vera H, López-Gay A, Peralta A, Solana-Solana M, Fernández A (2023) Gentrification and its association with health inequalities in Barcelona (2011–2017) [Original Research]. Front Environ Heal 2. https://doi.org/10.3389/fenvh.2023.1202149
He C, Huang Q, Bai X, Robinson DT, Shi P, Dou Y, Zhao B, Yan J, Zhang Q, Xu F, Daniell J (2021) A global analysis of the relationship between urbanization and fatalities in earthquake-prone areas. Int J Disaster Risk Sci 12(6):805–820. https://doi.org/10.1007/s13753-021-00385-z
Hosseini M, Farashi A, Khani A, Farhadinia MS (2019) Landscape connectivity for mammalian megafauna along the Iran-Turkmenistan-Afghanistan borderland. J Nat Conserv 52:125735. https://doi.org/10.1016/j.jnc.2019.125735
Hu M, Li C (2012) Design smart city based on 3S, internet of things, grid computing and cloud computing technology. In: Proceedings, Internet of Things: international workshop, IOT 2012, Changsha, China, August 17–19, 2012
Husgafvel R, Pajunen N, Virtanen K, Paavola I-L, Päällysaho M, Inkinen V, Heiskanen K, Dahl O, Ekroos A (2015) Social sustainability performance indicators—experiences from process industry. Int J Sustain Eng 8(1):14–25. https://doi.org/10.1080/19397038.2014.898711
Ibiwumi S, Evangelisca A (2022) Sustainable housing in developing countries: a reality or a mirage. In: Amjad A, Asaad A (eds) Sustainable housing. IntechOpen, pp Ch. 6. https://doi.org/10.5772/intechopen.99060
Iyer B (2020) Chapter 6—Cooperatives and the sustainable development goals. In: Altman M, Jensen A, Kurimoto A, Tulus R, Dongre Y, Jang S (eds) Waking the Asian Pacific co-operative potential. Academic Press, pp 59–70. https://doi.org/10.1016/B978-0-12-816666-6.00006-9
Janik A, Ryszko A, Szafraniec M (2020) Scientific landscape of smart and sustainable cities literature: a bibliometric analysis. Sustainability 12(3)
Javanmardi E, Liu S, Xie N (2023) Exploring the challenges to sustainable development from the perspective of grey systems theory. Systems 11(2)
Kleemann J, Inkoom JN, Thiel M, Shankar S, Lautenbach S, Fürst C (2017) Peri-urban land use pattern and its relation to land use planning in Ghana, West Africa. Landsc Urban Plann 165:280–294. https://doi.org/10.1016/j.landurbplan.2017.02.004
Kuddus MA, Tynan E, McBryde E (2020) Urbanization: a problem for the rich and the poor? Public Health Rev 41(1):1. https://doi.org/10.1186/s40985-019-0116-0
Kumar L, Chhogyel N, Gopalakrishnan T, Hasan MK, Jayasinghe SL, Kariyawasam CS, Kogo BK, Ratnayake S (2022) Chapter 4—Climate change and future of agri-food production. In: Bhat R (ed) Future foods. Academic Press, pp 49–79. https://doi.org/10.1016/B978-0-323-91001-9.00009-8
Lai S, Leone F, Zoppi C (2020) Spatial distribution of surface temperature and land cover: a study concerning sardinia, Italy. Sustainability 12(8)
Larimian T, Sadeghi A (2019) Measuring urban social sustainability: scale development and validation. Environ Plann b: Urban Anal City Sci 48(4):621–637. https://doi.org/10.1177/2399808319882950
Lin C, Zhao G, Yu C, Wu YJ (2019) Smart city development and residents' well-being. Sustainability 11(3). https://mdpi-res.com/d_attachment/sustainability/sustainability-11-00676/article_deploy/sustainability-11-00676.pdf?version=1548669506
Lin L, Yu S, Huang X, Tu X, Hong J (2021) Study on equilibrium relationship between government regulation and social environmental responsibility of energy enterprises. IOP Conf Ser: Earth Environ Sci

Lu Z, Wu Y, An S, Zhang Y, Zhu J (2022) The measurement and spatiotemporal evolution analysis of an ecological Niche for the high-quality development of resource-based cities in China. Sustainability 14(19)
Lu H, Shang Z, Ruan Y, Jiang L (2023) Study on urban expansion and population density changes based on the inverse S-shaped function. Sustainability 15(13)
Malla FA, Bandh SA (2023) Chapter 2—Biofuels and sustainable development goals. In: Hakeem KR, Bandh SA, Malla FA, Mehmood MA (eds) Environmental sustainability of biofuels. Elsevier, pp 13–26. https://doi.org/10.1016/B978-0-323-91159-7.00010-2
Meng X (2019) Rural–urban migration, urban poverty and inequality, and urbanization in the People's Republic of China. In: Jayanthakumaran K, Verma R, Wan G, Wilson E (eds) Internal migration, urbanization and poverty in Asia: dynamics and interrelationships. Springer Singapore, pp 77–107. https://doi.org/10.1007/978-981-13-1537-4_4
Mensah J (2019) Sustainable development: meaning, history, principles, pillars, and implications for human action: literature review. Cogent Soc Sci 5(1):1653531. https://doi.org/10.1080/23311886.2019.1653531
Moallemi EA, Malekpour S, Hadjikakou M, Raven R, Szetey K, Ningrum D, Dhiaulhaq A, Bryan BA (2020) Achieving the sustainable development goals requires transdisciplinary innovation at the local scale. One Earth 3(3):300–313. https://doi.org/10.1016/j.oneear.2020.08.006
Mueller L (2020) Chapter 63—1997 evolution of species' range—theory. In: Mueller L (ed) Conceptual breakthroughs in evolutionary ecology. Academic Press, pp 147–148. https://doi.org/10.1016/B978-0-12-816013-8.00063-6
Muhamad Khair NK, Lee KE, Mokhtar M (2020) Sustainable city and community empowerment through the implementation of community-based monitoring: a conceptual approach. Sustainability 12(22)
Munda G (2015) "Measuring sustainability": a multi-criterion framework. Environ Dev Sustain 7(1):117–134. https://doi.org/10.1007/s10668-003-4713-0
Musa AA, Malami SI, Alanazi F, Ounaies W, Alshammari M, Haruna SI (2023) Sustainable traffic management for smart cities using Internet-of-Things-oriented intelligent transportation systems (ITS): challenges and recommendations. Sustainability 15(13)
Mustafa MA, Mabhaudhi T, Avvari MV, Massawe F (2021) Chapter 2—Transition toward sustainable food systems: a holistic pathway toward sustainable development. In: Galanakis CM (ed) Food security and nutrition. Academic Press, pp 33–56. https://doi.org/10.1016/B978-0-12-820521-1.00002-2
Myers N (2019) Biodiversity and the precautionary principle. Ambio 22(2/3):74–79. http://www.jstor.org/stable/4314050
Najihah N, Indriastuti M, Suhendi C (2021) The effect of corporate social responsibility and environmental cost on financial performance. In: Proceedings of the 14th international conference on complex, intelligent and software intensive systems (CISIS-2020)
Olanrewaju ALA, Yeow TS, Tat LL (2015) Sustainable housing maintenance management model
Olanrewaju A, Tan SY, Abdul-Aziz AR (2018) Housing providers' insights on the benefits of sustainable affordable housing. Sustain Dev 26. https://doi.org/10.1002/sd.1854
Pasimeni MR, Valente D, Semeraro T, Petrosillo I, Zurlini G (2019) Anthropogenic landscapes. In: Fath B (ed) Encyclopedia of ecology, 2nd ed. Elsevier, pp 472–481. https://doi.org/10.1016/B978-0-12-409548-9.10602-5
Pastor R, Fraga A, López-Cózar L (2023) Interoperable, smart, and sustainable urban energy systems. Sustainability 15(18)
Paulin A (2019a) Chapter 2—Digitalized governance—an embezzled opportunity? In Paulin A (ed) Smart city governance. Elsevier, pp 39–60. https://doi.org/10.1016/B978-0-12-816224-8.00002-9
Paulin A (2019b) Governing the public apparatus through technology. In: Paulin A (ed) Smart city governance. Elsevier, pp 141–142. https://doi.org/10.1016/B978-0-12-816224-8.09991-X
Purvis B, Mao Y, Robinson D (2019) Three pillars of sustainability: in search of conceptual origins. Sustain Sci 14(3):681–695. https://doi.org/10.1007/s11625-018-0627-5

Qureshi AH, Alaloul WS, Musarat MA (2023) Sustainable development and urban design. In: Reference module in earth systems and environmental sciences. Elsevier. https://doi.org/10.1016/B978-0-323-93940-9.00066-9

Roggema R (2016) The future of sustainable urbanism: a redefinition. City, Territory Arch 3(1):22. https://doi.org/10.1186/s40410-016-0052-y

Rossi TF (2021) Travel demand forecasting: where are we and what are the emerging issues. In: Vickerman R (ed) International encyclopedia of transportation. Elsevier, pp 590–595. https://doi.org/10.1016/B978-0-08-102671-7.10388-4

Roy M (2021) Chapter 4—Sustainable innovation and corporate environmentalism. In: Roy M (ed) Sustainable development strategies. Butterworth-Heinemann, pp 91–131. https://doi.org/10.1016/B978-0-12-818920-7.00002-5

Setioningtyas WP, Illés CB, Dunay A, Hadi A, Wibowo TS (2022) Environmental economics and the SDGs: a review of their relationships and barriers. Sustainability 14(12)

Shah ZM (2023) Islamabad's vanishing forests. Dawn News. https://www.dawn.com/news/1743153#:~:text=The%20research%20shows%20that%20the,in%20the%20last%20two%20decades

Shahid N, Shah MA, Khan A, Maple C, Jeon G (2021) Towards greener smart cities and road traffic forecasting using air pollution data. Sustain Cities Soc 72:103062. https://doi.org/10.1016/j.scs.2021.103062

Shamsuddin S (2020) Resilience resistance: the challenges and implications of urban resilience implementation. National Library of Medicine (0264-2751 (Print)).

Siciliano G (2012) Urbanization strategies, rural development and land use changes in China: a multiple-level integrated assessment. Land Use Policy 29(1):165–178. https://doi.org/10.1016/j.landusepol.2011.06.003

Soliman AM, Soliman YA (2022) Exposing urban sustainability transitions: urban expansion in Alexandria, Egypt. Int J Urban Sustain Dev 14(1):33–55. https://doi.org/10.1080/19463138.2022.2056894

Soma H, Sukhwani V, Shaw R (2022) An approach to determining the linkage between livelihood assets and the housing conditions in urban slums of Dhaka. J Urban Manage 11(1):23–36. https://doi.org/10.1016/j.jum.2021.08.006

Staniškienė E, Stankevičiūtė Ž (2018) Social sustainability measurement framework: the case of employee perspective in a CSR-committed organisation. J Clean Prod 188:708–719. https://doi.org/10.1016/j.jclepro.2018.03.269

Torrens PM (2022) Smart and sentient retail high streets. Smart Cities 5(4):1670–1720. https://mdpi-res.com/d_attachment/smartcities/smartcities-05-00085/article_deploy/smartcities-05-00085-v3.pdf?version=1669856897

United Nation (2015a) Goal 11: make cities inclusive, safe, resilient and sustainable (Goal 11: Make cities inclusive, safe, resilient and sustainable, Issue). https://www.un.org/sustainabledevelopment/cities/?gclid=CjwKCAjw15eqBhBZEiwAbDomEp803eybjuhOB3y7bbS3W-BcfE71GbdPHtjRQMbfZ8D7rYsNMKNv0xoCqW8QAvD_BwE

United Nation (2015b) The 17 goals. United Nations. Retrieved 26–11–2023 from https://sdgs.un.org/goals

United Nation (2023) Sustainable lifestyles (Sustainable Lifestyles Issue). United Nations. https://www.unep.org/explore-topics/resource-efficiency/what-we-do/sustainable-lifestyles

UK TGo (2022). City of London cuts annual emissions by almost a third. SmartCitiesWorld. Retrieved 14–4–2023 from https://www.smartcitiesworld.net/climate-action/climate-action/city-of-london-cuts-annual-emissions-by-almost-a-third-8026

Wang M (2012) Chapter 40—Sustainable development. In Mannan S (ed) Lees' loss prevention in the process industries, 4th ed. Butterworth-Heinemann, pp 2507–2521. https://doi.org/10.1016/B978-0-12-397189-0.00040-9

Wang Z (2022) Research on smart city environment design and planning based on Internet of Things. J Sensors 2022:2348573. https://doi.org/10.1155/2022/2348573

Wang W, Zhang S, Zhang L, Liu Q (2020) Government subsidy policies and corporate social responsibility. IEEE Access 8:112814–112826

Wang P, Liu Z, Zhang L (2021) Sustainability of compact cities: a review of inter-building effect on building energy and solar energy use. Sustain Cities Soc 72:103035. https://doi.org/10.1016/j.scs.2021.103035

Wang X, Yang M, Park K, Um K-H, Kang M (2022) Social sustainability of a firm: orientation, practices, and performances. Int J Environ Res Public Health 19(20)

Wei G, He B-J, Sun P, Liu Y, Li R, Ouyang X, Luo K, Li S (2023). Evolutionary trends of urban expansion and its sustainable development: evidence from 80 representative cities in the belt and road initiative region. Cities 138:104353. https://doi.org/10.1016/j.cities.2023.104353

Xu R, Yue W, Wei F, Yang G, Chen Y, Pan K (2022) Inequality of public facilities between urban and rural areas and its driving factors in ten cities of China. Sci Rep 12(1):13244. https://doi.org/10.1038/s41598-022-17569-2

Yussif K, Dompreh EB, Gasparatos A (2023) Sustainability of urban expansion in Africa: a systematic literature review using the Drivers–Pressures–State–Impact–Responses (DPSIR) framework. Sustain Sci 18(3):1459–1479. https://doi.org/10.1007/s11625-022-01260-6

Zhang Q, Su S (2016) Determinants of urban expansion and their relative importance: a comparative analysis of 30 major metropolitans in China. Habitat Int 58:89–107. https://doi.org/10.1016/j.habitatint.2016.10.003

Zhang J, Lyu Y, Li Y, Geng Y (2022) Digital economy: an innovation driving factor for low-carbon development. Environ Impact Assess Rev 96:106821. https://doi.org/10.1016/j.eiar.2022.106821

Zhou W, Minnick MD, Cui C (2018) 2.08—GIS for natural resources (mineral, energy, and water). In: Huang B (ed) Comprehensive geographic information systems. Elsevier, pp 158–186. https://doi.org/10.1016/B978-0-12-409548-9.09643-3

Zhou R, Liu G, Zhang Y (2021) Sustainability evaluation and spatial heterogeneity of urban agglomerations: a China case study. Discov Sustain 2(1):1. https://doi.org/10.1007/s43621-021-00012-3

Zhou X, Moinuddin M, Renaud F, Barrett B, Xu J, Liang Q, Zhao J, Xia X, Bosher L, Huang S, Hoey T (2022) Development of an SDG interlinkages analysis model at the river basin scale: a case study in the Luanhe River Basin, China. Sustain Sci 17(4):1405–1433. https://doi.org/10.1007/s11625-021-01065-z

Living in Future World—Different People in Different Spaces

New Participatory Educational Spaces. Shared Urban Regeneration and the Role of Geography in Lifelong Learning of and About the Territory

Stefania Montebelli

Abstract The fight against land consumption and sustainable urban development is closely linked and represents the most important challenges of our time, in line with the EU's environmental ambition. Soil sealing, one of the most important causes of its consumption, derives precisely from urban expansion. The reuse of existing urban buildings would allow us to limit the further overbuilding of interstitial and rural green spaces due to urban dispersion, implementing the protection of the soil, an essential resource for human survival. Among the actions to combat, this serious problem is shared urban regeneration which, through the recovery of degraded areas, involves the urban community in a participatory project. From this perspective, the city can be considered an open-air laboratory, an educational and relational environment for continuous learning. In this context, knowledge of one's living space becomes crucial for conscious territoriality. In Italy, there are numerous shared planning initiatives in the territory, especially in neighbourhoods, and one could see in these spaces and participatory moments a new destination for the geographical discipline and the teaching of geography: spread knowledge of and about the territory outside conventional spaces, making this knowledge a tool for action for the shared planning of one's living spaces, educating active and aware citizenship. This is with the aim of spreading knowledge of and about the territory outside of conventional spaces, making geographical knowledge an action tool for the shared planning of common living spaces, and educating active and aware citizenship.

Keywords Shared urban regeneration · Educating city · Geographic lifelong learning · Education about the territory and active citizenship · Containment of land consumption

---

S. Montebelli (✉)
Department of Human Sciences, Guglielmo Marconi University, Rome, Italy
e-mail: s.montebelli@unimarconi.it

© The Author(s), under exclusive license to Springer Nature Switzerland AG 2025
D. Piróg and A. Świętek (eds.), *Future-Ready Geography*, Key Challenges in Geography, https://doi.org/10.1007/978-3-031-81268-2_10

Introduction

It is estimated that between 2018 and 2050, 2.5 billion more people worldwide will be living in the cities, and the inevitable urban growth will threaten the biotypes by contributing to increased global warming and catastrophic weather events, as well as straining food production through soil degradation and consumption (Decoville and Feltgen 2023: 1). Particularly on this last long-standing issue, in the context of the EU Soil Strategy for 2030, the European Commission presented a legislative proposal in July 2023 to monitor the state of EU soils with a view to making them healthy by 2050 (European Commission 2023a). The EU Green Deal aims to address soil pollution and consumption. This has led to funding for energy transition and promotion of alternative energy sources. Additionally, soil conversion projects are being promoted to supply the EU's highly urbanised areas. The depletion of soil, an essential resource for life and ecosystems, is closely linked to urbanisation and sealing. It is important to note that sustainable urban development policies have focused on addressing the acceleration of soil consumption and depletion resulting from urban expansion since the 1990s.[1] However, despite the use of Structural Funds to improve the administrative capacity of local authorities in order to "implement integrated urban development policies, also with a view to achieving the overall result of quality and sustainability of the built environment" (European Union 2007: 7), the urban sprawl of major European cities, as well as that of a growing number of cities around the world, has not stopped.

To counter the centrifugal impulse to conquer land outside the city, a centripetal impulse can be used to find consumed and artificial land within the city limits. This will help to build a sustainable and regenerated urban space. The urban tissue often contains spatial holes and traces of obsolete and disused buildings, which can be reconverted through urban regeneration.[2] This process, in contrast to the consumption of land that leads to the artificial creation of new rural spaces, works on the recovery and integration of the "lying" building in the urban context, becoming an opportunity for sustainable urban development. Turning the gaze inside the city, rather than looking for new ground, the regeneration of spatial voids without memory or function in the existing urban fabric represents a possibility for regenerating a new

[1] The promotion of sustainable urban development in the European Union includes: the Green Paper on the Urban Environment (1990), the Aalborg Charter (1994), the Framework of Actions for Sustainable Development (1998), the European Landscape Convention (2000), the Thematic Strategy for the Urban Environment (2004), the Leipzig Charter (2007), the Urban Agenda (2016), the Handbook of Sustainable Urban Development Strategies (2020), up to the "Lighthouse programme" and the EU Mission on smart and climate impact cities under Horizon 2020, renewed also in the period 2021–2027. It is specified that the Special Report "Smart cities Tangible solutions, but fragmentation challenges their wider adoption" (2023) emphasised the need for greater cooperation between the two programmes.

[2] Buildings defined as obsolete are those that, despite continuing to fulfil the function for which they were created, no longer respond to the new needs of the city; while disused buildings, defunctionalised of their original use, dissolve over time into urban interstitial spaces of decay, becoming semiophores awaiting re-signification (Assmann 2011).

urbanity to reinterpret the entire urban narrative. It is a process that produces a dense city, as opposed to the consumption of land, and an alternative to the expanded city.

It has to be said that urban regeneration is not only a process that triggers spatial reorganisation, but also activates cultural, economic, environmental, and social interventions to improve the lives of citizens and meet their long-term expectations. In this sense, the proposals and actions carried out in the territory are the result of a need for change expressed by the local context and the local community. In fact, urban regeneration can only be talked about when the rights to use space are multiplied by a heterogeneous public, when public and private spaces become a resource available to all citizens. In this way, urban regeneration becomes a mooring for empowerment processes based on political initiatives and the activation of local communities (Ostanel 2017: 7).

Involving the local community in the planning of shared living spaces, while taking into account their interests and needs for these spaces, can make urban regeneration a collaborative process. Communities should not passively accept the decisions of local government regarding public spaces but should play an active role in their design. This can lead to improvements in the quality and relevance of interventions in the area. However, to achieve consciousness, this process must be supported by lifelong learning of the territory. It is essential to train all social stakeholders to ensure sustainable and lasting actions on the territory. Therefore, it is fundamental to train local communities in the knowledge of their territory to intervene appropriately and respect an urban vision that links living spaces to the needs of the community, resolving any marginalisation. This is one possibility of the future role of geography and the importance of a "geographically based approach to learning" (Longworth and Osborne 2010: 369), through which local communities can be trained in the knowledge of their own territory in order to make their actions on it conscious and aware, in a process of continuous learning and collective self-education. This is a form of place-based education (PBE) where outdoor learning involving local communities is based on the experience of the territory and the places where people live. It is supported by educational networks and partners, and this education aims to return to the territory, for instance, through shared participation in participatory urban regeneration projects, for the benefit of its use and protection. A virtuous circle of intentions and participatory actions in which the territory is both the educational environment and the subject of learning: the "situated context" (Bruner 1996) to which educational methods are adapted (Bateson 1972), the formative laboratory in which social experience is put into practice and from which the active tendencies of each individual are derived (Dewey 1916). Collaborative urban regeneration can be a powerful tool for transforming a city according to the needs and actions of its citizens. To achieve this, education on conscious territoriality is essential.

Educating City: The Role of Territory and of the Geography

Urban regeneration is the process of improving living spaces through participatory and conscious collaboration among different social groups. It is an expression of the potential inherent in the concept of the Educating City. Animated by what Jacques Delors calls the "learning society", the Educating City is based on the possibilities of lifelong learning since: "in which everything affords an opportunity of learning and fulfilling one's potential", (Delors 1996: 38). Urban space thus becomes an educational resource by becoming a space of socialisation, i.e. of the intelligibility of common experiences that forge the individual's *attitude*[3], and a lived space by entering the emotional and experiential sphere of each individual (Lowenthal 1961). The city can be seen as a cultural space where individuals learn to be social beings through shared cultural experiences. The built environment of the city can be interpreted as a text composed of individual significations that shape the urban narrative and its communities. The city functions as an open-air school, educating the community on a daily basis. To be considered truly educational, the city must become an organism capable of activating virtuous and formative Lifelong Education[4] processes. This will empower its inhabitants to become active participants in shaping their reality. According to the Organisation for Economic Co-operation and Development (OECD), knowledge is not limited to predefined places or momentary episodes. Instead, it is an emanation of the concept of Innovative Learning Environments (ILE), which views knowledge as closely linked to an organic ecosystem that allows for mixed approaches and interdependent relationships. Learning evolves from this dense network (OECD 2013: 22, 23). In this way, the learning space should not be configured exclusively as neutral and closed within walls institutionally assigned to education, but as an environment open to complexity, relationality, and culturally and emotionally involving. In this sense, the city can be considered an organic ecosystem and open-air school par excellence of lifelong learning capable of creating possible networks of opportunities for universal access to essential educational resources (Reimer 1971).

The city is configured as both a learning city and an educating city, two sides of the same coin. On the one hand, the Educating City is a network of public and private actors in a community who learn to organise their sociality and living spaces

[3] Yi-Fu Tuan speaks of *attitude* in accordance with the concept of *cultural stance* that is determined on a "long succession of perceptions, that is, of experience", (Tuan 1974: 4). From this reciprocity between the individual and the context of life, both *topophlia* as: "the affective bond between people and place or setting", and a world view that: "is conceptualized experience. It is partly personal, largely social. It is an attitude or belief system", (Tuan 1974: 4).

[4] *Lifelong Education* covers the three forms of learning: formal, non-formal, and informal. The knowledge, skills, and abilities of formal education are acquired in educational institutions divided into the various school cycles and in higher education institutions that grant a vocational certificate or qualification. Non-formal learning can be conducted in various organisations with educational and training purposes outside the formal education and training circuit. Finally, Informal learning refers to: "activities carried out in every-day life, at work, at home and in leisure time, even without an intentional choice", (European Commission 2023b).

through participatory sharing. This is achieved through an interconnected educational network of lifelong and lifewide learning (UNESCO 2015: 8). On the other hand, the Educating City promotes lifelong learning through the natural surroundings and daily experiences, extending beyond the formal education system.

> Clearly, the concepts of the formative city and the learning city are intertwined and complementary, for territory can only become formative in terms of its value (material and symbolic) to the extent that the community experiences it and identifies itself in its spaces, thus helping to make it a relational, historical, and identity-forming environment (Augé, 1992). To be truly educational, the learning city must fulfil its primary conditions, including, as stated by UNESCO, the mobilisation and use of resources, the governance and participation of all stakeholders, and the strong will and political commitment of its citizens. (Montebelli 2023: 40)

In this interpretation of the Educating City, the territory can thus be considered as the diffuse environment of lifelong learning, of which it becomes the primary object of knowledge, the means, and the end through which to educate an active citizenship for a participatory planning of common spaces. A *knowledge society* based on sharing[5] is inseparable from the cultural context, but also from the reified context of the living space to which one must be trained in order to be able to interpret its value and consciously contribute to its transformation with respect to the ever-changing needs of living. Therefore, presenting the territory as an excellent educational environment, with its own urban fabric as a formative text, would be beneficial. The material reality of the territory can be considered as a geostructure, which refers to its geogram and the cultural significance of the territory's history (Raffestin 2007). A text, then, is made up of material structures capable of becoming referents, material signs of value that narrate the processes of territorialisation that have made it up over time, not without erasures and points of suspension. However, in order to read and interpret the territory, the community must be educated over time and through the various learning processes of lifelong education, which will have the task of nourishing the construction of identity that binds the community to its territory. This text defines a dialectic relationship between society and the territory, where the territory passes on knowledge to the community and the community participates in its construction. The text follows a logical structure with causal connections between statements and employs a formal register with no biased language or unnecessary jargon. The text adheres to style guides and citation consistency, and quotes are clearly marked. This concept is known as the Educating City, where education is reflected in the territory through community action. The language used is clear, objective, and free from grammatical errors, and technical terms are explained when first used. The content of the improved text is as close as possible to the source text, with no new aspects added. The community engages in participatory democracy to self-determine through conscious action on the territory. This leads to a territorialisation of the learning process (Montebelli 2023), where the territory is seen as a substrate

---

[5] On the concept of Knowledge Society, we refer in the first instance to the UNESCO text in *Towards Knowledge Society,* which also states that: "To remain human and liveable, knowledge societies will have to be societies of shared knowledge", (2005: 5).

for social self-representation and knowledge propagation, contributing to continuous collective learning. For the same reason, the Educating City is constantly evolving as a design workshop that trains active citizenship through awareness of its spaces, their meaning, and their transformation into participatory planning that adapts to changing times.

The city can serve as a laboratory for active citizenship. This creates new educational opportunities beyond formal learning that geography can cover. The goal is to train individuals in understanding the urban territory and contribute to the regeneration of its settlement and social fabric. From this perspective, one could consider the concept of learning organisations. These organisations could include companies, professional associations, schools, universities, and groups of people who are masters of the discipline. They could support learning about the territory in all its forms, contributing to social innovation in the name of inclusiveness and active participation in common making (Longworth 2006).

This is a reflection of the role of geography in citizenship education, which, as John Dewey explained, is inherent in its ability to relate natural facts and social events to their consequences in the local environment:

> Learning geography' means acquiring the ability to perceive the spatial, natural connections of an ordinary act (...). The variety of peoples and environments, their contrast with familiar scenes, provides infinite stimuli. The mind is moved by the monotony of the usual. And while local or domestic geography is the natural starting point in the reconstructive development of the natural environment, it is an intellectual starting point for moving into the unknown, not an end in itself. When not treated as a basis for accessing the vast world beyond, the study of domestic geography becomes as deadly as object lessons that simply summarise the properties of familiar objects. (Dewey 1916)

Geography provides an understanding of reality in its complex relationships between natural and human systems. An understanding that starts from the position and role that each individual plays in his or her daily life and, from there, in the world. In this sense, geography is a fundamental cultural tool in the education for democratic citizenship, expressed in a transcalar link from the local to the global.[6] Obviously, geography, as the science of territory, analyses how the perception of space and its shared representation are culturally constructed, entering into the specificities of the value system that marks some places and justifies their transformation. In fact, geography organises the knowledge of places by means of platforms, including the map, which allows spatial control and new forms of planning. It is undeniable that this has to do with the destiny of man and the planet, with the ways of living through the sharing of places and the use of resources (IGU 2016). Thus, the role of geography has an immense ethical as well as strategic value, closely linked to the design and educational perspective that determines, in cognition and transformation, the world in which one wants to live (Giorda 2021: 48). Therefore:

---

[6] In this sense, Geography can be considered as a: "translocal learning": learning centred on place, but not limited to this place. Translocal learning involves the continuous work of creating and developing connections between different sources, routes, and actors. [...] Urban learning is not simply spatially delimited in local places, but is instead relationally produced", (McFarlane 2011: 2).

> Geographic thinking is indeed a 'powerful thought' and can push people to take responsibility through knowledge that leads them to become more actively engaged citizens. This is what Edelson (2011, after Palacios et al. 2017) describe as 'geoliteracy' that in turn develops into 'geocapabilities'. Geographers are challenged to discover, or rediscover, the importance of the manifold and complex connections between nature and society and recognise that a more fundamental (and useful) understanding of landscapes. (Meadows 2020)

In this perspective, the International Geographical Union (IGU) describes the geographic knowledge as a conceptual and methodological tool to address the major global challenges of the twenty-first century: "helping people to think critically about sustainable living locally and globally and how to act accordingly" (IGU 2016: 10). In 2020, on the occasion of the International Year of Global Understanding, the IGU reiterated the importance of everyday actions and, in close connection with the UN SDGs, the impossibility of global change actions that do not start from the local dimension with a bottom-up process. This means that citizenship education must be promoted from the bottom up, with a daily education that involves people in the experience on and of the territory and encourages them to act together. An interconnection between different spatial and social levels that enables an understanding of the diversity of places and their identities, as well as their interconnections. Geography is a discipline that facilitates an understanding of interculturality. This understanding begins with living in a culturally and spatially diverse territory, which is linked to global destinies. Understanding one's place in the world starts with knowing one's own territory and being aware of the responsibility that daily actions carry for the common living space. This leads to a sense of multiple citizenship that combines the destiny of the local, national, and European territories, until we feel that the whole world is our common and shared home. The process of territorialisation of knowledge is crucial in transforming an individual's territory of life to make it functional for the needs of the community. This legal transcale highlights the importance of this process in regenerating the territory. Thus, geographic education transforms the territory into a laboratory of citizenship, serving as a hub of knowledge and a platform for the realisation of a shared vision in which the community expresses its values. The field of geography promotes territoriality and fosters awareness of the community's active role in shaping its living space. The project aims to discover the power of sharing and its effectiveness in practice through knowledge of the territory. This is achieved through territorial reification and its democratic, regenerative transformation.

Shared Design and Social Innovation: Urban Regeneration

Combined with a decline in urban population density and growth rate, urban sprawl is one of the most obvious and worrying changes in land use. The unplanned pattern of artificialisation that drives urban sprawl often creates new areas of development, particularly along major thoroughfares. This can cause environmental degradation and the loss of natural habitats that are converted to human use. Today, 80% of

Europe's land is artificial, with an increase of 6% between 2000 and 2018 (European Environment Agency 2023). More specifically, it can be said that: "The discontinuous urban fabric is the most important and dynamic artificial land use class in Europe" (Urbieta et al. 2019). Together with pollutant emissions, urban sprawl—with its artificialisation, sealing and progressive degradation—generates an unsustainable consumption of land (natural, semi-natural and agricultural), which, overexploited, is increasingly at risk in terms of health: "During urban expansion, a large amount of cultivated land, forests and shrubs are converted into urban land use, which leads to many environmental problems such as urban heat islands, increased carbon emissions and urban waterlogging, which pose a serious threat to local sustainable development (…)", (Huiyuan et al. 2023: 2). In the European Union, the comparative analysis of *land cover and land use*[7] shows that degradation is so significant that 60–70% of soils are considered unhealthy due to biodiversity loss, compaction, organic matter decline, salinisation, pollution and erosion (for the latter alone, between 2012 and 2018, more than 400 km^2 of soil were consumed per year on a net basis) (European Commission 2023a).

This implies not only environmental damage, but also considerable economic costs related to energy consumption in the commercial, transport, and residential sectors, with particular reference to public services and the construction of adequate infrastructure to connect functional living and working areas. The high environmental and economic price of urban expansion outwards has led to the adoption of a different logic that, by reversing the horizon and perspective towards the interior part of the city, finds in the regeneration of the existing built-up area an effective urban development strategy to counteract new land consumption, manipulating the face of a new territorialising rationality from the rejected reifications of the past. But that is not all. The process of urban regeneration is recognised as a multi-stakeholder one, in which interventions on the built and urban heritage can influence the development of cities at social, cultural, and environmental level. This is because, on the one hand, if urban regeneration interventions are implemented in the most deprived and peripheral areas, not only spatially, for the society that lives there, on the other hand, it should be considered that the urban regeneration process is inherent in the participation of the various social, public, and private parties in the planning phase. This collaboration aims to improve urban resilience by initiating territorial transformation processes that are decided in a participatory manner by various social actors and by listening to the needs of local communities. In this way, a process is set in motion that involves citizens, administrations, and public and private sectors in the design and implementation of living spaces. Obviously, such a virtuous process requires forms of metropolitan governance capable of supporting the territory and its community through shared structural policy choices and an integrated method of managing urban

[7] The term land cover defines the physical cover present on the earth's surface, including vegetation (natural or planted) and human constructions. The term land use refers to the resources, activities, and inputs that people undertake to produce, modify, or maintain some type of land cover. Therefore, with respect to these definitions, land use establishes a direct link between land cover and the actions of people in their environment (Bui and Mucsi 2021).

complexity.[8] In this sense, urban regeneration is a process of empowerment that activates local communities to participate in political initiatives on the territory. It can therefore be said that at the heart of urban regeneration policies, in addition to the reified asset to be revitalised, is the cultural process that drives the transformation of urban spaces by directing public and private investment (Paddison and Miles 2020). Thus, urban regeneration, participatory cultural processes, and alternative governance actively coexist to promote social innovation. An urban social innovation that implies the transformation not only of physical space, but also of social relations, consolidating the identity and cultural ties that bind the community to the place. This means that social innovation should not be seen as a fixed concept, but rather as a "rule" in the making, both in theory and in practice (Moulaert et al. 2013). This rule is shaped in relation to the specificity of the place and the community that inhabits it, that is, the 'situation' in which the innovation takes place:

> In other words, the 'placing' of urban social innovation reveals neither a generic geographic context nor a particular kind of place; instead, its placing unveils the ways in which the 'urban' is differently imagined and put to work. The concern becomes, then, less one of figuring out what is specifically "urban" and what is "non-urban" about social innovation, but instead one of how the urban is enacted and differently enters into the composition of urban social innovation. (McFarlane et al. 2023)

Thus, urban regeneration can be considered as an opportunity for social innovation, given its characteristics of participation and consultation between the different stakeholders who become active participants in the process. This is because the decision-making process of the urban regeneration project is not only the prerogative of the public subject, which continues to play a clearly defined central role, but it is extended to the final users of the asset, private stakeholders, owners, trade, and citizens' associations, who can actively participate in the various stages of the process. On the contrary, it can be said that the involvement of citizens in spatial planning, although it can take different forms, as the Arnstein[9] scale also shows, must be considered fundamental in order to counteract the exclusive interrelationship between government and large interest groups. The consultation of the various

---

[8] The collaboration between the various social, public, and private actors, which drives urban regeneration, constitutes the basis of a process that would not stop at the city sphere alone, but could lead to a territorial rescaling with which to implement a re-territorialisation and re-definition in a trans-scale logic of the various levels of government in order to make the actions of territorial governance at different scales coordinated and synergic. The process of rescaling creates forms of reticular capital whereby territorial governance is structured in multiple levels of management (multilevel governance). This means that the metropolitan city moves towards the capital city of the macro-region, the municipality moves towards intercommunality, the region towards the macro-region, the state towards neo-federal Europe, etc. An approach that would allow a continuous confrontation between the local dimension and the supra-local dimension strengthening territorial cohesion and local development.

[9] Sherry Arnstein, in a 1969 study, analysed the different forms of citizen participation in US planning processes. The scholar devised one of the different degrees of participation in decision-making power with respect to which one can go through six intermediate levels, from a "manipulated" and non-active citizen participation, through which the necessary public support can be obtained, to a "citizen control" that manages the entire planning, drafting and management programme (Arnstein 1969).

social partners, both in the planning phase and in the implementation and management of the regenerated asset that is part of the urban fabric, is therefore the guarantor of the preservation of a representative democracy, which is: "is the cornerstone of modern governance. It provides a framework for citizen participation, political decision-making, and accountability. In the face of increasing volatility and rapid social change" (European Conference of Presidents 2023). The urban regeneration project holds all these elements together by fostering a process of attachment, empowerment, and awareness from the community of the reality and places with which they identify. This is why such a project cannot be imposed from above, with a top-down process from the institutions, because it is strongly linked to the context where it develops and is in no way transferable to other environments and other physical and social realities. The development of the urban regeneration project begins with the creation of an exploratory, operational, and finally interpretative knowledge: not only a problem-solving activity, but one that involves all the cognitive abilities of the community. It begins with the recognition of needs and the identification of problems; it moves on to the exploration of the space and the cultures that inhabit it and have inhabited it in the past, with the recognition of its lost identity and its current identity; the consequent awareness of the relativity of the point of view on the same object and the possible conflicts of interests, aspirations, and visions on it; the confrontation with the unavoidable dimensions of the economy and the legal system, probably also with inflexible actors such as institutions of different scales; the elaboration of solutions and their social negotiation on the basis of collective needs. Here, then, urban design contributes to the individual, social, experiential, operational, and creative learning process, inextricably linked to reality, everyday life, emotionality, logic, sharing, motivation, sociality, and the city as a whole. Thus:

> Within a commons framing, acquiring knowledge needs to be recast as not simply enabling individuals, but rather as connecting individuals to one another and intergenerationally to the common knowledge resources of humanity. A commoning paradigm requires attention to the collective ways knowledge is accessed, used, and created. The knowledge pillar supporting education would then need to be oriented towards *learning to study, inquire and co-construct together*. This revision would highlight the social dimensions of learning, as well as the diverse and networked dimensions of knowledge. (UNESCO 2023)

This gives life to an urban dimension animated by a relationship of reciprocity between urbs and civitas, which finds in the territory its potential for social innovation and its laboratory for collective action. That is why, in this context, urban planning can guide and strengthen local self-organisation processes, based on a vision of the collective subject, involving the physical as well as the social context of the territory. For this reason, urban regeneration must be seen as a multidimensional process, materially manifested in the functional and architectural aspect of the inhabited space, but whose implementation is based on various components of urban life that give it meaning and significance. It can therefore be said that the urban experience of participatory collaboration between residents and urban technicians in the construction of the living space becomes a fundamental part of living and being in the world. And if dwelling is the construction of places by taking care of them (Heidegger 1971), in a continuous negotiation and appropriation of urban space, then education in territorial awareness cannot be separated from learning as social participation:

> Participation here refers (...) to a more encompassing process of being active participants in the practices of social communities and constructing identities in relation to these communities. (...). Such participation shapes not only what we do, but also who we are and how we interpret what we do. (...) participation as a process of learning and of knowing. (Wenger 1998: 4)

In this perspective, the active participation of the community in urban projects of common spaces can be considered the educational tool for citizenship training, in a continuous learning that finds in the city its testing ground, but also its open classroom in which to learn to know, to learn to do, to learn to be, to learn to live together (Delors 1996).

Participatory Urban Regeneration: Some Italian Examples

In Italy, there are a number of projects for participatory urban regeneration, which in fact represent new educational spaces for the geographical discipline, thanks to which different groups of players can be trained to know the territory, with the aim of rethinking shared urban spaces. A cooperative and educational rethinking, through lifelong learning, to which geography can undoubtedly contribute, aiming at the involvement of the urban governance network, of governmental and non-governmental community players, capable of creating relationships of trust and collaborative practices. A network of social stakeholders that focuses on the territory, starting from the needs that only those who live there can clearly see, such as everyday needs. From this point of view, the role of the citizen is fundamental, as co-author of the transformation of the territory of life, which, as space of interactivity, becomes a theatre of social action (Turri 2001) and a symbol of collective unity (Mumford 1938: 475), as well as a widespread formative environment that educates by inhabiting and transforming it.

A representative example of what is being done in Italy in the field of participatory urban regeneration[10] is being promoted by the Order of Architects of Rome, together with the town all, to create a training and production centre for urban design, that aims to promote the culture of community participation in urban regeneration and redevelopment processes. In fact, at the end of February 2024, the project, and the administrative documentation for the construction of the first Metropolitan Urban Centre of Rome were delivered, whose main intention will be precisely the involvement of citizens in urban planning processes and public policies (Fig. 1).

[10] The following examples enrich the picture of Italian initiatives concerning participatory urban regeneration projects that have been analysed in two previous contributions (Montebelli 2021, 2023) concerning the active citizenship workshops Idee per una città accogliente promoted by the University of Sassari and the path towards the Urban Innovation Plan, with Neighbourhood Laboratories, launched by the Municipality of Bologna.

Fig. 1 The metropolitan urban center will be built inside the ITIS Galileo Galilei building in Viale Manzoni in Rome and the works will be sponsored by Roma S.p.a., metropolitan city of Rome, order of landscape and conservation architects of Rome and its province—APP, resources for Rome

A space that will deal with the urban territory as a whole, as well as in its individual municipal parts, with the help of citizen participation in urban transformation processes. Training in territorial knowledge will be promoted through multi-purpose and multimedia sharing. It will be possible to design and submit ideas, to take part in events that will encourage comparative synergy, but also cooperation between the various social stakeholders with national and international bodies and institutions, thus promoting dialogue and scientific and cultural debate between all the social actors involved in these processes. As stated in the tender for the planning and implementation of the Metropolitan Centre of Rome, it will be aimed at telling the story of the city and, in particular, of its current and future urban transformations.

It must be said that Roma Capitale has long been aware of the importance of citizen participation in urban regeneration projects and of the need for territorial education to make this participation conscious. What until the early 2000s was defined as "public communication" has in fact been gradually consolidated, initiating a dialogue with residents, involving them more and more in decision-making processes and thus rethinking the role and value of democratic participation. This has been made possible by supporting the various forms of consultation through which the local community can be directly involved in urban development, now increasingly facilitated by the possibilities offered by digital technology, without, however, replacing public debate. A commitment that the Capitoline Assembly has envisaged since 1994 and that, in January 2018,[11] was reshaped within the conceptual framework of popular participation, giving it an important role with the introduction of new institutions such as the participatory balance, the proposed referendum, and the electronic popular petition. In particular, the Participatory Balance is a fundamental tool for the construction of e-democracy and for the management of economic resources to be invested in interventions of urban redevelopment of places proposed by citizens at the municipal and neighbourhood level. Precisely in 2018, a pilot project was launched in the VIII district of Rome, involving residents and users of the city, i.e. students and workers. A project that, from 2019, will give way to the experimentation of other urban initiatives. The *Bilancio Partecipativo Capitolino* (Capital Participatory Balance Sheet), although not the first in Italy, is undoubtedly the most articulated, due to the size of the territory involved, so much so that it required a set of rules and regulations to involve all segments of the citizenry, including the less digitised, through public meetings organised in the area. However, the telematic platform is of fundamental importance as it allows citizens to upload their proposals and express their preferences on the proposed projects. It should also be remembered that, in order to promote the culture of knowledge of the territory and participation in its transformation, the project for the Register of Active Citizenship and Civic Networks of Roma Capitale was presented in February 2024, which promotes cooperation between associations, committees, voluntary organisations, third sector organisations, civic bodies, and active citizens organised in formal and informal groups.

Another example of the importance of citizen participation in urban regeneration processes is that of the Tuscan municipality of Rosignano Marittimo, in the province of Livorno, which, in collaboration with the Department of Educational Sciences of the University of Roma Tre, has launched the *Rosignano Comunità Domani* project between the end of 2020 and the beginning of 2022, as part of the work on the new Structural Plan. It is worth remembering that the Structural Plan is an important tool through which the local administration defines medium and long-term development strategies for the territory and identifies the instruments to implement them. Territorial planning, therefore, which makes explicit the vision with which it intends to achieve sustainable development, capable of integrating, improving, and regenerating urban areas and their potential. As established by the Regional Law 65/2014 *Norms for the Government of the Territory,* the formation process of the Structural Plan has provided a formative and cognitive process of the structural components[12] of

[11] We refer to the paragraph amended by Resolution of the Capitoline Assembly of the Statute of Roma Capitale, No. 5 of 30 January 2018, Chapter II, "Popular participation and protection of civic rights".

[12] The structural components referred to are environmental, landscape, economic, settlement, infrastructure, etc.

the municipality necessary to define the development strategies to be implemented. What is most interesting in this discussion is that the standard establishes that this cognitive process consists of moments of listening and participation of the citizenship. This objective was pursued through: public meetings to present the framework of the characteristics of the territory and the possible shared scenario of its transformation; the involvement of schools to understand how the territory is perceived by young citizens; interviews with economic and social actors active in the community to understand the feasibility of the proposed strategic lines. The indications and observations resulting from these moments finally allowed the definition of the Structural Plan. One of these moments was the *Rosignano Comunità Domani* project, which had the task of analysing the degree of territoriality of the municipality, as an indispensable element of territorial cohesion, in order to assess the soundness of the social tissue, at the basis of any future vision and possible territorial development (Burgalassi 2023). It should be stressed that the individual's sense of belonging is nourished and translated into a sense of community, that is, in the "a sense of similarity with others, an acknowledged interdependence with others, a willingness to maintain this interdependence by giving to others or doing for them what one expects of them, a sense of being part of a larger, reliable and stable structure" (Sarason 1974: 157). A sense of belonging and community that motivates common action, as a realisation of identity cohesion. The aim of the *Rosignano Comunità Domani* project was precisely to study the sense of belonging to the territory and, therefore, the degree of cohesion of its community, in order then to encourage the participation of citizens in a democratic project of territorial regeneration.

The research used a two-stage mixed methods approach. Firstly, the social networks and ties of the community fabric were analysed through a survey. Citizens were able to contribute to this phase by filling in a web self-administered questionnaires—WSAQ—an online tool to simplify the data collection process. The questions posed by the questionnaire ranged from the meaning given by the individual to the community, the link with the territory, the impact of the community on the daily life, the willingness to participate in the society, the possible future prospects that would guarantee the belonging the community and dealing with the functioning of the relationships created during the coronavirus pandemic emergency. Out of the 1200 citizens involved, 760 duly filled in the survey. Then, in the second stage, a qualitative survey was carried out by a semi-structured interview, using the Delphi method, to a specific section of the community knowing the local reality (no more than 10 units). The interviews were designed to reveal the community's perceptions of their ability to influence the future of local society, and how. The questions asked therefore concerned: the characteristics of the area where to focus for the future of the local community, the medium-term prospects for the community tissue, the degree of social cohesion and the motivating elements, the proposals to promote a sense of belonging to the area and to the community. The first stage of the analysis carried out showed that the social tissue of Rosignano has a considerable degree of internal cohesion, since it is rooted in the territory in terms of values and facts, as demonstrated by the civic commitment of the majority of the inhabitants, although some nuances were highlighted depending on the different areas of the inhabited territory.

This was largely confirmed by the interviews conducted in the second stage of the project to specific identified witnesses. Such interviews revealed not only a difference in the perception of the sense of belonging to the community between the parts of the coastal area that are well served and the hillside area that has suffered from an impoverishment of services and trade, but also between the younger, more fragile, and the older, more deeply rooted segments of the population. The conclusions of the project have therefore highlighted the need to promote a sense of social cohesion throughout the territory as an essential basis for any possible territorial project that seeks to promote common well-being. A sense of cohesion and belonging that can be fostered through education for conscious citizenship, with training sessions capable of stimulating discussion on common territorial issues. This is why it is important to promote participatory processes in order to strengthen the community's attachment to the territory through a sense of identity that is actively expressed in the common project for the development and evolution of its living space.

Conclusion

Through these examples of participatory urban regeneration, this article highlights the value of involving all social partners in shared space projects. It emphasises that the serious problem of land consumption caused by unstoppable urban sprawl should not be considered generically "global", but rather requires a transcale approach. Thus, a network logic of interdependence, inspired by a hierarchy of adaptive cycles that operate at different scales and communicate with each other, defines the behaviour of the system as a whole (Walker and Salt 2006). Resilience actions to "global" problems can be implemented and extended from contexts at large and very large scales, including the smallest neighbourhoods that make up a city. This article examines the importance of the empowerment process that activates local communities to be involved in political initiatives on the territory, such as participatory urban regeneration. This approach favours the valorisation of the existing common territorial heritage and counteracts the consumption of new land at the infinitesimal level of the urban. Training in territorial knowledge is crucial for initiating participatory democracy projects and supporting urban regeneration proposals. Thus, democratically designing the city can become the main tool for dealing with the complexity of everyday living space, through which we learn to find ourselves as a community: "(…) the territory becomes educational in its complicated and therefore generates training experiences" (Valentino and Lutzoni 2020, p. 5).

From this perspective, educating for citizenship means understanding the city as a combination of civitas and urbs, and overcoming the traditionally divisive nature-artifice dichotomy. Every corner of the city can serve as a classroom for incidental education, where the environment is used for formative experiences that allow for experimentation with autonomy and sharing in social life (Ward 2018). Designing for the adaptive reuse of an existing building or space is a possibility for a participatory workshop when its previous function becomes obsolete, as well as a practice for combating land consumption[13] reason why: "As cities are both sources of these problems and innovation areas for dealing with them, and as they are laboratories, in theory and in practice, for resilience, solution options relating to urbanization and the built environment become crucial", (Gültekin 2021: 5). The urban territory is considered the ultimate educational space. It has the ability to rewrite itself over time through the shared action of its citizens, annexing and repurposing its parts. This increases the complexity of its meaning because:

> Le territoire, tout surchargé qu'il est de traces et de lectures passées en force, ressemble plutôt à un palimpseste. Pour mettre en place de nouveaux équipements, pour exploiter plus rationnellement certaines terres, il est souvent indispensable d'en modifier la substance de façon irréversible. Mais le territoire n'est pas un emballage perdu ni un produit de consommation qui se remplace. Chacun est unique, d'où la nécessité de " recycler ", de gratter une fois encore (mais si possible avec le plus grand soin) le vieux texte que les hommes ont inscrit sur l'irremplaçable matériau des sols, afin d'en déposer un nouveau, qui réponde aux nécessités d'aujourd'hui avant d'être abrogé à son tour. (Corboz 1983, p. 35)

The process of urban regeneration involves a continuous rewriting of the territorial narrative. If truly participated in, it can be considered a laboratory of self-organisation and democratic innovation that educates towards an inclusive society and personal development (Delors 1996, p. 80). The process of configuring living is inseparable from the knowledge of the territory and the appropriation of one's role within the space and community:

> "To inhabit" refers to a kind of having (habere) that permits us a radical access to material objects and allows us to treat these objects as extensions of our own body. To approach inhabitation in this way means to be able no longer to make such a radical distinction between flesh and matter, between bodies and mere things. Bodily existence floods over into things, appropriates them, infuses them with the breath of life, draws them into the sphere of daily projects and concerns. A fully inhabited world is at the same time also a fully embodied world. (...) To enter and finally to come to inhabit a house or city means to come to assume a certain stance, to surrender to a certain style of acting upon and of experiencing the surrounding world. (Jager 1985: 219)

Geography, with its specialised knowledge, can make an incisive contribution to the process of appropriation through the promotion of lifelong learning based on the education of the spatial dimension that underlies the construction and orientation of identity, both individual and collective. Orientation refers not only to the ability to move within a given space but also to the knowledge of how to navigate within a cultural context of symbols that imbue places with meaning. The educational challenge of geography is to make individuals aware of their surroundings and their role within it, motivating responsible actions on the territory they inhabit. Living means taking responsibility for actively engaging with the territory and becoming co-authors of its narration. In this sense, the territory is the result of actions induced

[13] In this sense, it should be recalled that the EU objective of soil protection also includes the rehabilitation and reuse of the built environment, which is why one-third of the resources provided by the *Next Generation EU* have been earmarked for climate protection actions and urban regeneration with policies for the renovation of existing buildings. Especially the European Regional Development Fund (ERDF) in the period 2021–2027 deals with funding for sustainable urban development.

by a syntagmatic actor at every level (Raffestin 1980) through active and positive territoriality. This capability constructs one's own territory of life in processual terms, manifesting itself in social practices and inclusive logics. (Banini 2019: 217).

From this perspective, geography should enhance the development of spatial intelligence, enabling individuals to become social actors based on their territoriality, which is a sense of identity that underlies every territorial action. To disseminate these skills, geography can explore new educational spaces, such as those offered by participatory urban regeneration. This can go beyond institutional classrooms and include non-formal learning opportunities, allowing communities to become aware of the space of coexistence before becoming active. This approach can help create a global understanding starting from the local level. Territorial awareness is fundamental for participatory democracy and requires geographical literacy to understand the cultural meaning inscribed in the urban landscape. Geography educates us to read the narrative of territorialisation and represents the Earth as humanity's dwelling through the activities, forms, and effects of inhabitation. On the other hand, if geographical knowledge were to lose its capacity to narrate the habitation of the Earth, it would be reduced to: "a veritable pile of intellectual junk: the height of a mountain here, the course of a river there, the quantity of roof tiles produced in this town, the tonnage of ships in that, the boundary of a county, the capital of a state. The earth as man's home is humanising and unifying; the earth seen as a miscellany of facts is dispersed and imaginatively inert (Dewey 1916)".

References

Arnstein SR (1969) A ladder of citizen participation. J Am Plann Assoc 35(4):216–224

Assmann A (2011) Cultural memory and western civilization. Functions, media, archives. Cambridge University Press, Cambridge

Banini T (2019) Geografie culturali. FrancoAngeli, Milano

Bateson G (1972) Steps to an ecology of mind. Chandler Publishing Company, London

Bruner J (1996) The culture of education. Harvard University Press, Harvard. https://doi.org/10.4159/9780674251083

Bui DH, Mucsi L (2021) From land cover map to land use map: a combined pixel-based and object-based approach using multi-temporal Landsat data, a random forest classifier, and decision rules. Remote Sens 13(9):1700. https://doi.org/10.3390/rs13091700

Burgalassi M (2023) Una comunità locale tra presente e futuro. Roma Tre Press, Roma, Senso di appartenenza e coesione sociale nel territorio di Rosignano

Corboz A (1983) Le territoire comme palimpseste. Diogéne:14–35. https://www.marcellinbarthassat.ch/files/le_territoire_comme_palimpseste.pdf

Decoville A, Feltgen V (2023) Clarifying the EU objective of no net land take: a necessity to avoid the cure being worse than the disease. Land Use Policy 131:1–8. https://doi.org/10.1016/j.landusepol.2023.106722

Delors J (1996) Learning: the treasure within; report to UNESCO of the international commission on education for the twenty-first century. UNESCO. https://doi.org/10.7788/ijbe.1996.24.1.253

Dewey J (1916) Democracy and education. In: An introduction to the philosophy of education. The Free Press, New York

European Commission (2023a) Commission Staff working document impact assessment report. Annexes. Accompanying the proposal for a Directive of the European Parliament and of the Council on Soil Monitoring and Resilience (Soil Monitoring Law). https://environment.ec.europa.eu/system/files/2023-07/IMPACT%20ASSESSMENT%20REPORT_ANNEXES_SWD_2023_417_part4.pdf

European Commission (2023b) Validation of non-formal and informal learning. https://eurydice.eacea.ec.europa.eu/national-education-systems/italy/validation-non-formal-and-informal-learning

European Conference of Presidents of Parliament (2023) Conférence européenne des Présidents et Présidents de Parlement. https://rm.coe.int/background-paper-theme-2-the-challenges-to-representative-democracy-in/1680abe22a

European Environment Agency (2023) Land use. https://www.eea.europa.eu/en/topics/in-depth/land-use

European Union (2007) Leipzig charter on sustainable European Cities. https://territorialagenda.eu/wp-content/uploads/leipzig_charter_2007.pdf

Giorda C (2021) Conoscenza geografica e cittadinanza. Un progetto per il territorio. In: Giorda C, Puttilli M (eds) Educare al territorio, educare il territorio. Geografia per la formazione. Carocci, Roma, pp 45–54

Gültekin Y (2021) Strategies to improve urban energy efficiency for urban resilience. Mater Sci Eng 1203:1–10. https://doi.org/10.1088/1757-899X/1203/2/022020

Heidegger M (1971) Poetry, language, thought. Harper Colophon Books, New York

Huiyuan L, Zhengyong S, Yanling R, Linlin J (2023) Study on urban expansion and population density changes based on the inverse S-shaped function. Sustainability 15(10464). https://doi.org/10.3390/su151310464

International Geographical Union—IGU (2016) International charter or geographical education. Commission on Geographical Education. https://www.igu-cge.org/wp-content/uploads/2019/03/IGU_2016_eng_ver25Feb2019.pdf

Jager B (1985) Body, house and city: the intertwining's of embodiment, inhabitation and civilization. In: Seamon D, Mugerauer R (ed) Dwelling, place and environment. Towards a phenomenology of person and world. Martinus Nijhoff, Dordrecht, pp 215–225

Longworth N (2006) Learning cities, learning regions, learning communities. In: Lifelong learning and local government. Routledge, London

Longworth N, Osborne M (2010) Perspectives on learning cities and regions: policies, practice and participation. NIACE

Lowenthal D (1961) Geography, experience and imagination: toward a geographical epistemology. Ann Assoc Am Geogr 1961(51):241–260

Mumford L (1938) The culture of cities. A harvest/HBJ book, Orlando. https://monoskop.org/images/5/5a/Mumford_Lewis_The_Culture_of_Cities.pdf

McFarlane C (2011) Learning the city. Knowledge and translocal assemblage. Wiley-Blackwell, Hoboken.

McFarlane C, Langley P, Painter J, Lewis S, Vradis A (2023) Interrogating 'urban social innovation': relationality and urban change in Berlin. Urban Geogr 44(2):337–357. https://doi.org/10.1080/02723638.2021.2003586

Meadows ME (2020) Geography education for sustainable development. Geogr Sustain 1(1):88–92. https://doi.org/10.1016/j.geosus.2020.02.001

Montebelli S. Il territorio come spazio della creatività sociale. Un esempio di laboratorio per una "città accogliente". Educ Sci Soc (1):40–55. https://doi.org/10.3280/ess1-2021oa10030

Montebelli S (2023) The role of the territory in an educating city: two Italian laboratory of participatory urban planning. Eur J Geogr 14(4):1–9. https://doi.org/10.48088/ejg.s.mon.14.4.037.045

Moulaert F, MacCallum D, Hillier J (2013) Social innovation: intuition, precept, concept, theory and practice. In: The international handbook on social innovation, pp 13–24. https://doi.org/10.4337/9781849809986.00011

OECD (2013) Innovative Learning Environment. OECD Publishing. https://doi.org/10.1787/9789264203488-en
Ostanel E (2017) Spazi fuori dal comune: rigenerare, includere, innovare. FrancoAngeli, Milano
Paddison R, Miles RS (eds) (2020) Culture-led urban regeneration. Routledge, London
Raffestin C (1980) Pour une géographie du pouvoir. Paris, èd. LITEC
Raffestin C (2007) Il paesaggio come strumento progettuale per il territorio. http://www.osservatoriodelpaesaggio.org/AZIONI%20svolte/2007/Premiazione%20Bando%20Paesaggio%20(Passerano%20Marmorito%2027%2010%2007)/Prolusione%20Prof.%20Claude%20Raffestin%202.htm
Reimer E (1971) School is dead: an essay on alternatives in education. Doubleday & Company, New York
Sarason SB (1974) The psychological sense of community: prospects for a community psychology. Jossey-Bass, San Francisco
Tuan YF (1974) Topophilia. A study of environmental perception, attitudes, and values. Columbia University Press, New York
Turri E (2001) Il paesaggio come teatro. Dal territorio vissuto al territorio rappresentato. Marsilio editore, Venezia.
UNESCO (2005) Toward knowledge societies. UNESCO World Report. Conde-sur-Noireau
UNESCO (2015) Global network of learning cities. Guiding documents. UNESCO Institute for Lifelong Learning, Germany
UNESCO (2023) Reworking four pillars of education to sustain the commons. https://www.unesco.org/en/articles/reworking-four-pillars-education-sustain-commons
Urbieta P, Fernandez E, Ramos L, Mèndez-Martinez G, Bento R (2019) A land-cover based urban dispersion indicator suitable for highly dispersed, discontinuously artificialized territories: the case of continental Portugal. Land Use Policy 85:92–103. https://doi.org/10.1016/j.landusepol.2019.03.048
Valentino M, Lutzoni L (2020) Among territorial project and architectural design. City, Territ Arch (7):1–12. https://doi.org/10.1186/s40410-020-00115-w
Walker B, Salt D (2006) Resilience thinking: sustaining ecosystems and people in a changing world. Island Press, London
Ward C (2018) L'educazione incidentale. Elèuthera, Milano
Wenger E (1998) Communities of practice: learning, meaning, and identity. Cambridge University Press, Cambridge. https://doi.org/10.1017/CBO9780511803932

Mapping the Future Landscape: Ghettoization Patterns of Roma Population in Northwestern Bulgaria

Nadezhda Ilieva, Kaloyan Tsvetkov, Boris Kazakov, and Dimitar Enikiev

Abstract In the wake of Bulgaria's post-communist era, the chapter delves into the complexities of structural changes, unveiling new socio-spatial polarization models. A notable trend emerges as the Roma population increasingly relocates to urban areas often "hidden" for the authorities, establishing quarters within existing neighborhoods or forming entirely new ones. Focused on towns and cities in Northwestern Bulgaria, the study utilizes orthophoto images and QGIS tools to identify and map Roma neighborhoods and track their dynamics in the last decades. By using urban and socio-spatial indicators like ethnic homogeneity, location, ethnic surroundings, demographic trends, and formation time, the research categorizes these structures. This comprehensive analysis provides insights into the multifaceted nature of Roma neighborhoods and aligns with the overarching theme of future-ready geography. By dynamically visualizing the evolving landscape, the study contributes to preparing geographical science and education for future challenges, emphasizing the adaptability of geographical methodologies in understanding, and responding to societal changes in Northwestern Bulgaria. The chapter underscores the importance of examining broader social processes and historical patterns in understanding how policies impact Roma populations and contribute to marginalization. The findings emphasize the need for new policies to address and mitigate ongoing segregation impacting Bulgaria's Roma population.

Keywords Ghettoization · Urban Roma · Spatial polarization · Social segregation

N. Ilieva · B. Kazakov
National Institute of Geophysics, Geodesy and Geography, Bulgarian Academy of Sciences, Akad. Georgi Bonchev 3 Str., Sofia 1113, Bulgaria
e-mail: nadeto.ilieva@abv.bg

K. Tsvetkov (✉)
Faculty of Geology and Geography, Sofia University St. Kliment Ohridski, Tsar Osvoboditel 15 Blvd., Sofia 1000, Bulgaria
e-mail: ktsvetkov@gea.uni-sofia.bg

D. Enikiev
The University of Library Studies and Information Technologies, 7-Mi Kilometar, Boulevard "Tsarigradsko Shose" 119, 1784 Sofia, Bulgaria

© The Author(s), under exclusive license to Springer Nature Switzerland AG 2025

D. Piróg and A. Świętek (eds.), *Future-Ready Geography*, Key Challenges in Geography, https://doi.org/10.1007/978-3-031-81268-2_11

Introduction

In the twenty-first century, often referred to as "Century of the City" (United Nations 2017), we witness unprecedented urbanization rates, presenting a spectrum of challenges and opportunities. This urban transformation, particularly in post-socialist contexts, includes the increasing segregation of Roma communities, a phenomenon garnering significant scholarly attention against the backdrop of Central and Eastern Europe's depopulation trends. The emergent socio-economic disparities and spatial segregation are among the foremost challenges confronting European urban centers in the foreseeable future.

This research underscores the intensification of spatial segregation driven by growing social polarization, a trend transcending the confines of Western cities. Contemporary studies spotlight new patterns of socio-spatial polarization in post-socialist urban landscapes, rooted in income disparities and ethnic divisions. Despite socialist-era policies aimed at fostering equality and reducing segregation through social housing initiatives, the Roma populace remains markedly isolated within spatial enclaves. Settlements dominated by Roma, spanning from Bulgaria to the Czech Republic, are integral to the ethnic composition of numerous post-socialist cities. These communities have seen substantial growth, leading to a plethora of social and infrastructural challenges. This study seeks to elucidate the dynamics of these changes and their implications for urban spaces, proposing an analytical methodology for examining ghettoization patterns among the Roma population in Northwestern Bulgaria. The objective is to delineate and categorize distinct types of ghettoized zones within cities, considering their unique characteristics and impact on societal development and urban integration. Through this analysis, the study aims to unravel the complex mechanisms of ghettoization, laying the groundwork for formulating social integration strategies tailored to the specific needs of the Roma community. In doing so, this chapter contributes to the broader discourse on fostering more inclusive and sustainable urban environments.

Urban spaces are inherently shaped by their social structures. Many scholars posits, the urban fabric can be read as a map that records societal structures. The nexus between social processes and spatial configurations has increasingly captivated academic inquiry, propelling the evolution of sophisticated theoretical frameworks regarding their interplay. Durkheim et al. (1963) contend that space is intrinsically linked to social organization and is interpreted variably across cultures. Urban space theory pioneers, such as Harvey (1973, 1982) and Castells (1972), conceptualize urban spaces as mirrors of capital accumulation and social reproduction processes. Massey (2005) extends this discourse, viewing economic development through the prism of spatial labor division and its ensuing effects on urban morphology.

On the other hand, Harvey (2006) and Simmel (1997) argue that space is not merely a physical or geometric entity but a social construct. Every entity occupies a unique spatial position, forming what Bourdieu (1989) describes as a "system of relations" or, as Massey (2005) terms, a "product of relations." This socio-spatial dynamic underpins the notions of the "dual city" and the "divided city," wherein urban spaces reflect and perpetuate societal divisions and segregation. Amid escalating social polarization, urban landscapes become increasingly fragmented, mirroring profound societal schisms. This phenomenon highlights the critical importance of examining spatial segregation as a gauge of social integration or disintegration at a global scale.

Numerous academics (Peach 1975; Massey and Denton 1993, among others) argue that spatial segregation patterns reflect the processes of integration and that areas populated predominantly by lower social status groups hinder the possibility of social advancement for their inhabitants. Haussermann (2005) characterizes segregation as "the projection of social structures into space," while Francini (2013) suggests that the emergence of "ethnic enclaves" within cities can compromise the social, cultural, and spatial coherence of urban environments. In the United States, segregation is typically analyzed through metrics related to the ethnic or racial backgrounds of individuals. Conversely, in Europe, factors like socio-economic status, education levels, and housing quality are more significant indicators of segregation (Ilieva et al. 2020).

Theoretical Framework and Methodology

Theoretical Framework

Spatial segregation can refer to various territorial divisions such as ghettos, ethnic enclaves, urban ghettoized structures, and slums, known by different names worldwide like "Favelas" in Latin America and "Shanty Towns" in Africa and others. This phenomenon occurs across different scales within a city, from wider areas to specific buildings. The term "ghetto," while Central to spatial segregation discussions, is often used imprecisely, leading to confusion and misuse. A true ghetto is characterized by the forced segregation of a group along ethno-social lines, differing from voluntary social clustering or poverty-stricken areas without forced separation. Other scholars define ghettos by demographics, with some focusing on race, others on class, and still others on ethnic dominance. Marcuse (2005) categorizes urban spatial groupings into ghettos, enclaves, and citadels, based on the nature of their formation and the dynamics with surrounding populations. Knox and Pinch (2010) discuss the relationship between majority and minority populations, detailing the nuances between colonies, enclaves, and ghettos, where colonies and enclaves represent voluntary groupings with the former being a transient phenomenon and the latter more enduring.

In the Czech academic discourse, "socially excluded areas" denote segregated spaces housing marginalized groups, which could range in scale and exhibit various degrees of separation. The concept of ghettoization describes the creation of urban spaces that segregate ethnic or racial minorities due to poverty or other social factors.

Some researchers focus on the integration policies for Roma communities, exploring strategies to address and mitigate the impacts of segregation.

This study adopts the term "ghettoized urban structures" (GUSs) as defined by Asenov (2018) to refer to parts of urban space, such as districts or neighborhoods, often enclosed and marked by: (a) inadequate or absent street and building infrastructure, including electricity and social services; (b) a primarily segregated population along ethnic, racial, social, or cultural lines; (c) community relations often governed by pre-modern social norms; and (d) the existence of a distinct subculture. These areas are significant components of the urban fabric, impacting the city's functionality due to their complex, intertwined economic, social, and environmental issues, recognized through distinct morphological features.

Ghettoized urban structures can range in size from several neighborhoods to a single district or part thereof and have shown considerable expansion and emergence of new areas in recent decades. The challenges these communities face are consistent with other similarly marginalized areas across the country. Criteria for identifying GUSs include ethnocultural composition, socio-economic indicators (e.g., unemployment, poverty rates, informal employment, low education levels), and infrastructural factors (e.g., poor housing quality, inadequate public transport, and high rates of illegal constructions).

Addressing the multifaceted issues within GUSs and formulating effective policies require accurate geospatial data integrated with additional information. Despite the visibility of their problems, the expansion of GUSs, especially Roma communities, often remains unnoticed by the public, further complicated by their unauthorized settlements on both public and private lands. This lack of visibility impedes monitoring their spatial development and addressing their needs comprehensively.

Data and Methods

Gathering and analyzing information about informal Roma settlements is a significant challenge. It's essential to collect spatial data on these areas and integrate it with other information to evaluate their status, develop improvement strategies, and prevent their expansion. Many Roma ghettoized urban structures are built illegally on public and private lands, making them hard to monitor. Effective policy-making and integration efforts require comprehensive spatial and demographic data obtained through various methods:

- **Census Data**: Utilizes demographic and socio-economic data from population censuses. While offering broad insights, it lacks detail at smaller geographical levels and might not capture the dynamics within Roma communities due to its infrequency and lack of detail at finer spatial scales.
- **Remote Sensing**: Uses drones or satellite images for mapping, offering high accuracy but may be constrained by costs and technical needs.

- **Mixed Data Collection**: Employs both qualitative and quantitative methods, providing insights from community engagement but is resource-intensive.
- **Visual Satellite Imagery Analysis**: Involves interpreting satellite images to identify GUS based on physical layout. While efficient, it may overlook the socio-economic context of these areas.
- **Machine Learning and GIS**: Automates the mapping process using image classification algorithms, providing speed and scalability but requiring high-quality images and technical expertise.

Each method has its benefits and limitations, influenced by factors like cost, data availability, and project scope. A combined approach could yield the most comprehensive understanding of Roma GUS, supporting informed policy-making and enhancing living conditions.

This research uses mixed data collection from different sources and machine learning in order to track the dynamic of GUS. By using orthophoto images, we can monitor changes in Roma neighborhoods since the early 2000s, highlighting the lack of comprehensive data on these communities in Bulgaria. Many Roma settlements lack proper planning, infrastructure, or legal recognition. These issues remain largely unaddressed by local authorities due to the settlements' "invisibility" in official records. The study utilizes also other sources of information to gather data such as official plans and strategies, governmental reports, cadastral data, interview with local authorities in order to document and analyze the evolution of these areas, noting new developments and expansions. Additionally, in-depth interviews for Lom municipality help to capture contemporary demographic shifts, despite national and local data shortages. For mapping buildings and infrastructure, the Mapflow plugin for QGIS was used, allowing for the digitalization of structures from the images. This data is further analyzed.

The study develops a nuanced slum identification framework, integrating a three-tiered ontology that outlines the morphology of slums at the environs, settlement, and object levels Fig. 1, following the structure proposed by Kohli et al. (2012). It emphasizes local customization to reflect specific slum characteristics, incorporating 27 indicators across these levels with varying emphasis.

These indicators, each bearing its distinct significance, encompass various facets of neighborhood infrastructure, quality of life, and diverse social indices. Such a comprehensive approach necessitates the classification of indicators based not on their individual values but on their categorical types, ensuring the applicability of the study across different neighborhoods and localities.

The proposed methodology differentiates between primary and secondary indicators—10 main (primary) indicators highlight core segregated area traits, while 17 additional (secondary) indicators offer supplementary context, see Table 1.

The evaluation scores each indicator from 1 to 3 to quantify spatial segregation levels, balancing the importance between primary and secondary indicators through a weighted formula. The analytical framework categorizes indicators into two primary

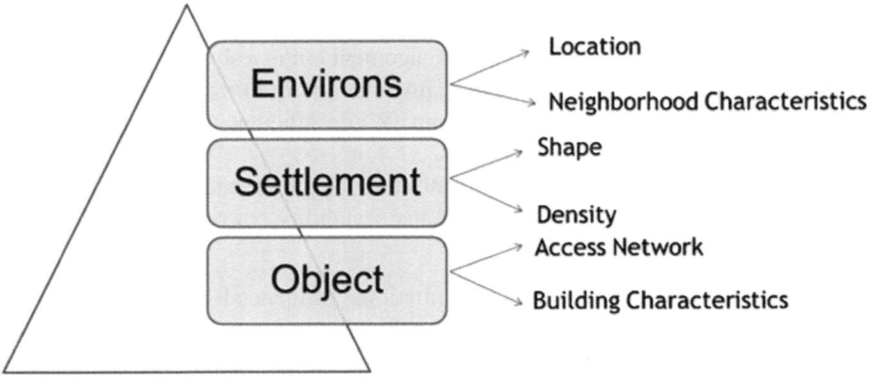

Fig. 1 Six general indicators categorized to form a hierarchy to represent concepts at three spatial levels (Kohli et al. 2012)

groups: Primary indicators (I_p) and secondary indicators (I_s), with the latter constituting approximately 60% of the indicators to reflect their broader scope. The evaluation employs a three-point scale alongside Boolean scoring for certain indicators, assigning extreme values consistent with the three-point scale methodology. Each indicator value represents a variable (x), integral to the subsequent analysis.

Given the pivotal role of primary indicators (I_p) in the overall assessment, they are accorded a weight (L_p) of 75 out of a possible 100 units. Conversely, secondary indicators (I_s) are allocated a weight (L_s) of 25 units, thereby achieving a balanced representation between the two indicator categories.

The modification of each indicator entails the multiplication of its weighting factor by the variable value (x), followed by the addition of the variable value (x) to the computed result. The formula devised seeks to elucidate the segregation level (Sl), calculated by dividing the aggregate of all modified indicators (S_{sum}) by their maximal potential score (S_{max}), subsequently multiplied by 100. This methodology ensures a nuanced understanding of spatial segregation, facilitating the derivation of actionable insights for policy formulation and urban planning.

The calculation implements a straightforward mathematical method which is given in Eq. 1.

Equation 1:

$$S_{\text{Sum}} = \sum I_p\forall + \sum I_s\forall \rightarrow S_l = \frac{S_{\text{sum}}}{S_{\max}} * 100 \tag{1}$$

$\sum I_p\forall$—Total sum of all primary indicators
$\sum I_s\forall$—Total sum of all secondary indicators
$I_p\forall = \left[\left(I_p(x) * L_p\right) + I_p(x)\right]$—Value formed of primary indicators
$I_s\forall = [(I_s(x) * L_s) + I_s(x)]$—Value formed of secondary indicators.

Table 1 Main indicators used to calculate the index of spatial segregation

	Level	Indicator
Main indicators	Level 1	**Location in the urban space**
		Access to school
		School segregation
		Access to kindergarten
		Access to public transport
		Proximity to anthropogenic hazards and potential pollution sources
	Level 2	**Access to tap water**
		Population density
	Level 3	**Residential area per inhabitant**
		Share of buildings with no ownership documents
Additional indicator	Level 1	*Access to emergency medical care*
		Access to outpatient care
		Access to public parks
		Natural hazard risk
	Level 2	*Road surface type*
		Proximity to major transport axes
		Sewerage
		Waste collection
		Green areas for public use
		Children's playground
		Built-up area dynamics
		Spatial expansion
		Morphology of the urban space
		Open space for public use
		Population dynamics
		Housing and sectoral policies
	Level 3	*Built-up area density*

The formula aims to derive a segregation level (Sl) by dividing the sum obtained from all (∀) modified indicators (S_{sum}) and their maximum possible score ($S_{\max}$), multiplied by 100, see Table 2.

Final results of the index may be three categories of segregation, see Table 3.

The approach facilitates converting qualitative observations into quantitative assessments, adaptable for analyzing different neighborhoods. Segregation levels are calculated by adjusting indicator values with weighting factors, aiming for a balanced evaluation between primary and secondary indicators. Population dynamics, density,

Table 2 Description of the parameters and variables

Parameter	Meaning	Description
Segment level	Sl	Level of segregation
Segment sum	S_{sum}	Final sum of all indicators
Segment max	S_{max}	Maximum sum of all indicators
Prime index	I_p	Primary indicators
Secondary index	I_s	Secondary indicators
Primary load	L_p	Weight to a main criterion with a constant value 0.75
Secondary load	L_s	Weight to a secondary criterion with a constant value 0.25
For each one	$\forall$	For all indicators
Variable	(x)	**Variable values** (x): 3—highly segregated/boolean value "YES"/ 2—partially segregated 1—low level of segregation/boolean value "NO"/

Table 3 Categories of segregation based on the aggregation of indicators

Range	Level of segregation
0.66–1	Definitive features of segregation are exhibited
0.33–0.659	Certain segregation features are exhibited
0–0.329	Some, to no segregation features are exhibited

built-up area changes, and proximity measures to various urban features are calculated using a combination of election data, national statistical institute (NSI) census data, Google Earth Pro imagery, and Geographic Information System (GIS) tools, ensuring a comprehensive spatial analysis of segregated neighborhoods.

Study Area

The study area covers only cities and towns in the municipalities of three districts—Vidin, Montana and Vratsa. They are part of the Bulgaria's Northwestern region. which is one of the poorest in the EU and also one of the most affected by depopulation processes in the country Fig. 2. The study area includes in total 32 municipalities, respectively, 11 in Vidin, 11 in Montana and 10 in Vratsa district. Of all centers of municipalities in the three districts, ghettoized urban structures have been formed in 12 of them. They are the subject of the study in this book chapter. Small GUS formed in villages and hamlets are excluded from the study.

The total number of studied GUS is 16 as for Lom there are 4 GUSs, and in Montana two GUS. Montana and Vidin according to demographic statistics in 2022, are the largest in terms of population, see Table 4.

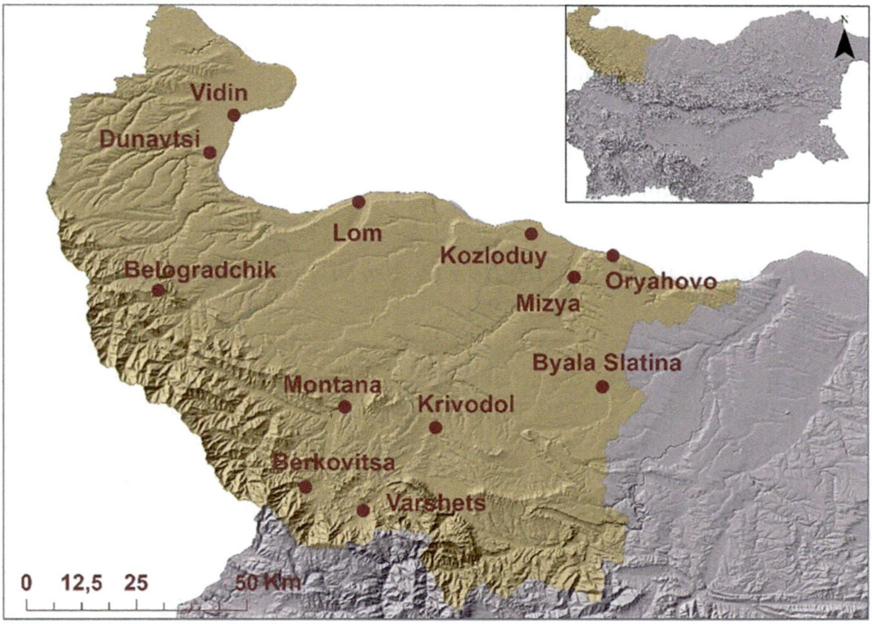

Fig. 2 Study area covers twelve centers of municipalities in three districts—Vidin, Montana and Vratsa

Table 4 Population number (National Statistical Institute 2023)

Settlement	District	Population number 2022
Montana	Montana	36,633
Vidin	Vidin	34,797
Lom	Montana	16,297
Kozloduy	Vratsa	10,916
Berkovitsa	Montana	9945
Byala Slatina	Vratsa	8975
Varshets	Montana	5308
Belogradchik	Vidin	3964
Oryahovo	Vratsa	3787
Krivodol	Vratsa	2600
Miziya	Vratsa	2373
Dunavtsi	Vidin	1693

The smallest towns are Krivodol, Miziya and Dunavtsi—all of them below 3 thousand people. The urban spatial structure showcases a diverse range of Roma neighborhoods within city limits, each distinct in its location and consequentially, in its degree of isolation. This separation is manifested not just physically but socially as well. The urban morphology of these areas reveals various Roma neighborhoods, each marked by unique traits that epitomize the most distinctive spatial configurations of ghettoized Roma communities across the country.

Results and Discussion

Description of Roma Population in Northwestern Bulgaria

As of September 7, 2021, Bulgarian ethnic group constitute 5,118,494 individuals, making up 84.6% of those who responded to the question of ethnicity in the census. This is a slight decrease of 0.2% from 2011, according to the Ethnocultural Characteristics' report the national statistical institute in Bulgaria (2022). The Roma ethnic group is ranked as the third largest in Bulgaria, with the census data revealing a count of approximately 325,000 individuals in 2011, which has seen a decrease to 266,000 by the year 2021.

Considering that ethnicity is only recorded during censuses and is self-reported by individuals, notable inaccuracies emerge. Significant gaps exist between the number of Roma identified in official census figures and those reported by scientific and empirical studies. Some scholars estimates suggest that the Roma population in Bulgaria could be as high as 720,000 (Ilieva and Bardarov 2020). This discrepancy is partly due to the phenomenon of "prefigured ethnic identity," where individuals may identify with an ethnicity other than Roma. Additionally, the notable proportion of respondents who did not declare their ethnicity in the 2021 census—amounting to 7.2% of the total counted—further complicates the accurate representation of ethnic distributions.

In recent years, approximately 16–18% of Bulgaria's Roma population has been concentrated in the Northwestern region. Post-World War II, this concentration diminished to 11% by the last census in 2011. The decrease is attributed to growth rates lower than the national average and higher increases in other areas, particularly the Southwest, Southeast, and South Central regions. Northwestern Bulgaria has consistently had the highest relative share of Roma. Until the early 1990s, the proportion remained stable at above 3%, but by 1992 it had nearly doubled to 6.2%, continuing to rise in the following years. By 2011, it reached 8.9%, twice the national average. This increase is partly due to the high negative rates of natural and migration growth within the Bulgarian population, accentuating the intense depopulation processes in the region, exacerbated by economic challenges.

Historically, from 1887, when the Roma population in Northwestern Bulgaria was recorded at 9213, it increased nearly fivefold to 46,349 by 2001. Due to increased emigration mobility and migration to other parts of the country since the twenty-first century's start, the number decreased to 35,434 in 2011.

By the early 1990s, over half of the Roma population in Northwestern Bulgaria was located in eight municipalities, each with more than 1000 Roma residents. This pattern remained consistent until the early 1990s. The municipalities include major cities like Vratsa, Vidin, and Montana, as well as Danube municipalities such as Lom, Oryahovo, Valchedrum, alongside Byala Slatina and Krivodol. In contrast, municipalities in the Western Balkan Mountains like Boynitsa, Kula, Makresh, and others, housed less than 5% of the region's Roma, indicating a minimal Roma presence.

Significant growth in the Roma population by 1992 was observed particularly in Lom and Vidin (each with over 6000 people), and Montana and Byala Slatina (each with over 3000 people). These municipalities alone comprised about half of the Roma population in Northwestern Bulgaria. Notably, Kozloduy and Berkovitsa also saw significant increases. Including municipalities with previously recorded high Roma counts, such as Valchedram and Vratsa, over 80% of the region's Roma population is concentrated in these areas.

The early 1990s marked a period where most municipalities had a Roma population share exceeding 3%, with exceptions in the mountainous Balkan region (Stara planina). Some municipalities had even higher relative shares, exceeding 4% and 5%, and places like Brusartsi and Valchedram surpassed 6%. Since the 1990s, there's been a notable increase in the relative share of the Roma population, including in mountainous municipalities. By 2011, many municipalities reported more than 15% Roma population, including Lom, Valchedram, and others.

Comparing the early records from 1887, where the Roma was distributed across 233 settlements, to the census in 2011 showing an expanded distribution to 331 settlements, illustrates significant changes in the territorial distribution of the Roma population in Northwestern Bulgaria.

Segregation Index

Based on the collected data we can categorize the Spatial Segregation Index (SSI) data for the 16 ghettoized urban structures.

The category of high segregation (index > 0.66) includes the GUS with the highest SSI values, indicating a high level of spatial segregation. The neighborhoods in this category are experiencing pronounced segregation, with strong distinctions in socio-economic and ethnocultural characteristics compared to the rest of the city. These areas have the highest segregation scores, implying intense spatial segregation with significant socio-economic and cultural disparities. The results indicate significant socio-spatial disparities in Northwestern Bulgaria's Roma GUS. These findings align with Bardarov and Tsvetkov's (2020) observation that historical marginalization and insufficient access to basic services exacerbate the challenges of segregation. The most spatially segregate neighborhood is "Kosharnik" in Montana with score 0.91, see Fig. 3.

There are six GUSs in Montana district and two in Vidin district which have definitive features of segregation.

- Montana (Kosharnik)—Montana District (0.91)

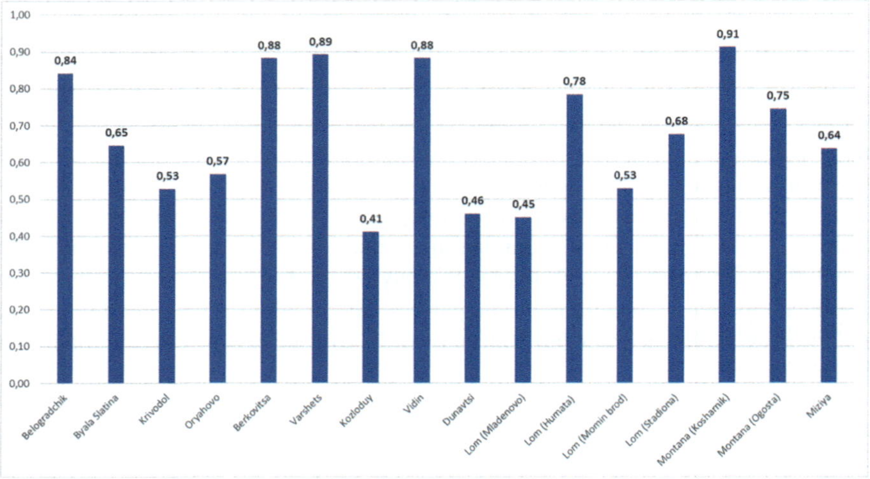

Fig. 3 Value of the segregation index by neighborhoods

- Varshets—Montana District (0.89)
- Berkovitsa—Montana District (0.88)
- Vidin—Vidin District (0.88)
- Belogradchik—Vidin District (0.84)
- Lom (Humata)—Montana District (0.78)
- Montana (Ogosta)—Montana District (0.75)
- Lom (Stadiona)—Montana District (0.68).

Analyzing the number of Roma in the city of Montana reveals a trend in their increasing numbers. The Roma community is concentrated in two neighborhoods "Kosharnik" and "Ogosta." According to 2011 data, the number of self-identified Roma is 3055, and in 2021—3152. According to data from local government representatives, the actual number of Roma living in the city is around 5000. The "Kosharnik" district is located in the northern part of the city, which is both larger in area and population, see Fig. 4. It was formed in the 1970s after the construction of a ceramics factory. The Roma originally settled in the space between "Nov zhivot" and "Kamchia" streets, but now occupy territories to the south-west and north-east. There are also parts of the neighborhood inhabited by both Roma and Bulgarian ethnic groups especially between the "Nov zhivot" str. and "Dimitar Eriiski" str. By 2009, 528 residential buildings existed in the Roma neighborhood, which increased by 82 in the period 2009–2014. After 2014, new 176 residential buildings appeared. The total area of the "Kosharnik" GUS is 0.55 km^2. Approximately 2600 people live there in 2023.

The second GUS in the city, "Ogosta," is situated in the city center, surrounded by the main stadium, the Montana–Berkovitsa railway line, two major boulevards, and the Ogosta River. Its segregation index is 0.75. The number of dwelling units counted by QGIS algorithms is 289. A distinctive characteristic of this neighborhood

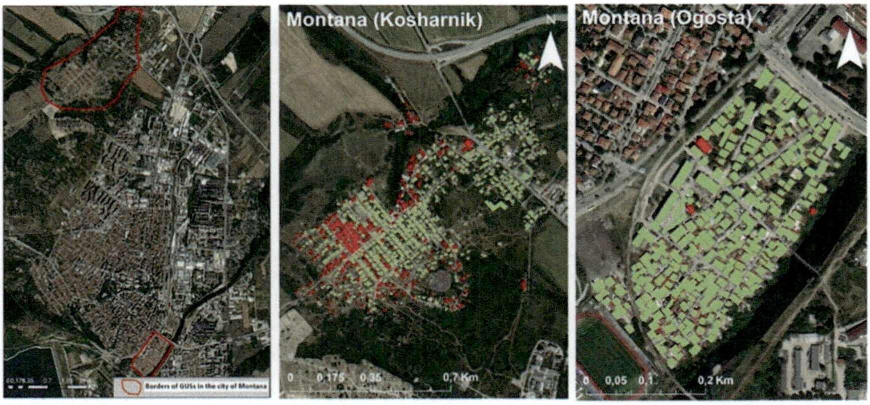

Fig. 4 GUS of "Kosharnik" neighborhood has the highest index of spatial segregation—0.91

is its spatial limitation and very high population density. The population density is estimated to be about 10,000 inhabitants per km^2. Practically "Ogosta" GUS cannot grow any more, as there are almost no free spaces around it. The majority of the houses (98.6%) were built before 2009. This neighborhood has lower index as it has access to some very important infrastructure objects—hospital school, kindergarten, public transport bus stops, etc., see Fig. 5.

One of the highest index is calculated for Vidin's "Nov pat" GUS. This neighborhood is the largest by size and population in Vidin district. The demographic analysis indicates a growing Roma population mostly with illegal buildings on the

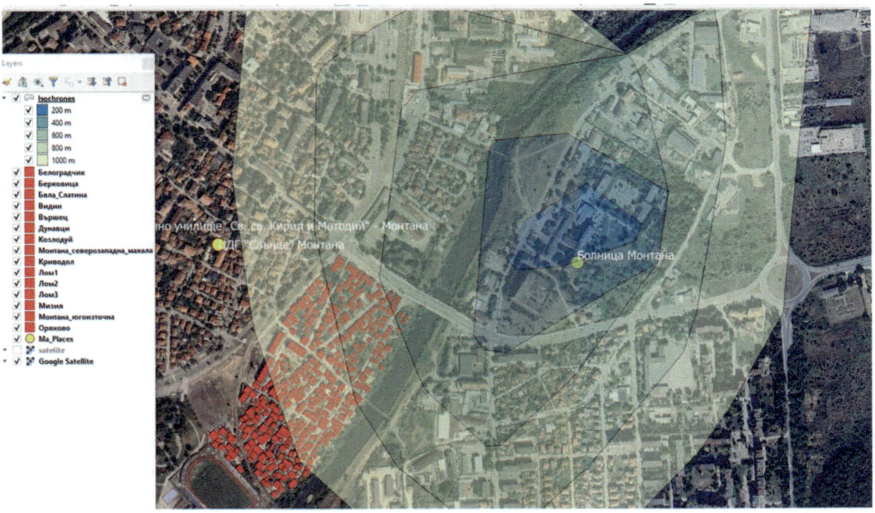

Fig. 5 Distance from the city main hospital to "Ogosta" GUS

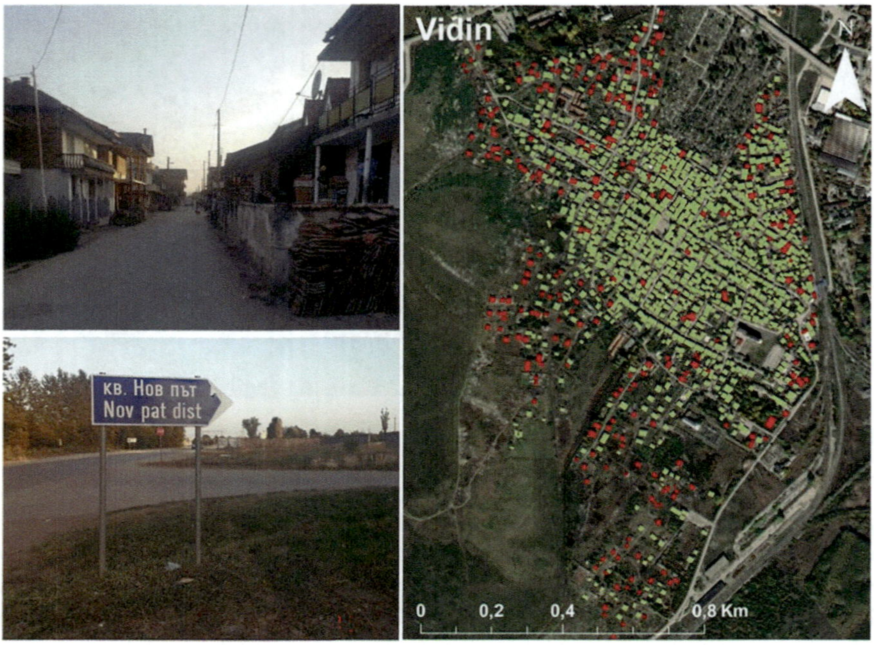

Fig. 6 Neighborhood "Nov pat" is the largest GUS in Vidin district

Western edge. Originally established between 1936 and 1938, this area was built on former Turkish and Roma burial grounds. Initially called "Bonchova Mahala" after Mayor Barni Bonchev, who facilitated relocation for families from the flood-prone "Bara Mahala," the district's name has undergone several changes, from "Boyanova Mahala" to "Boyan Chonos," after a local revolutionary leader, before settling on "Nov pat" (Stoyanova 2017).

Over time, the neighborhood has seen considerable growth with new residential developments, see Fig. 6.

Currently, about three-quarters of Vidin's Roma residents live locally, while the remainder has moved primarily to countries like Germany, England, Italy, France, and the Netherlands. Despite this significant diaspora, the local Roma community continues to expand its presence within Vidin.

Local officials estimate Vidin's Roma population at approximately 5000 individuals, making up about 11% of the city's total populace. Positioned in the north part, near the Central Vidin Cemetery and flanked by industrial zones to the northwest and east, the district's boundaries are least defined to the south and southwest, where new construction gradually encroaches on the city's ring road. This Roma quarter has significantly expanded in area, growing by 13% over the last decade. By 2022, the neighborhood included 1426 homes, with 291 of these constructed in the past ten years, representing 20.4% of the local housing stock, primarily on the neighborhood's outskirts.

In the second group of municipalities, a certain segregation features are exhibited and the SSI is **between 0.33 and 0.66**. This middle range indicates a moderate level of segregation. The GUS in this category may have some attributes that lead to segregation, but they're not as acute or as numerous as those in the first category. Based on the collected data, eight GUSs from Montana and Vratsa districts are part of this category. There is only one from Vidin municipality in this category.

- Byala Slatina—Vratsa District (0.65)
- Miziya—Vratsa District (0.64)
- Oryahovo—Vratsa District (0.57)
- Krivodol—Vratsa District (0.53)
- Lom (Momin brod)—Montana District (0.53)
- Dunavtsi—Vidin District (0.46)
- Lom (Mladenovo)—Montana District (0.45)
- Kozloduy—Vratsa District (0.41).

These areas have moderate SSI scores, which may reflect some social, economic, and spatial divisions, but not as intense as those in the high segregation group. The lower scores suggest that while some segregation may exist, it is less pervasive, and the differences between these areas and the rest of the urban fabric might be less pronounced.

Towns such as Kozloduy and Dunavtsi have one of the lowest score in the region. On of the main problems in Kozloduy is the shortage of council housing in the making it difficult to provide housing for families in need. Roma people are often categorized predominantly as at-risk groups. Currently, there are no mechanisms to stimulate investment in this direction, which limits the capacity of Kozloduy municipality to implement an adequate social housing policy. Currently, the following problems exist:

- The occupation of a single housing building by many families;
- Low environmental protection culture and accumulation of waste;
- Provision of funds for the repair and maintenance of housing and poor management of the same by their tenants;
- Shortage of council housing for families in need.

The percentage of illegal housing in Kozloduy is one of the lowest in the region. It is about 3% of all houses. At the same time the segregation exist in the following indicator—access to kindergarten, see Fig. 7.

Index below 0.33 suggests a lower level of segregation. The GUS here would have fewer characteristics that lead to segregation, and they would be less pronounced compared to the other two categories. Based on our research no municipalities within this category exist in Northwestern Bulgaria.

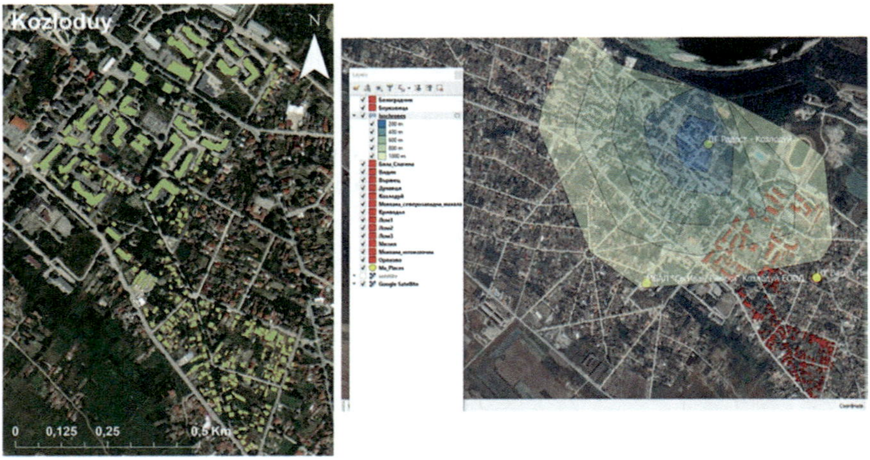

Fig. 7 GUS in Kozloduy and the distance from the closest kindergarten

Accessibility of Essential Services in Roma Ghettoized Structures in Northwestern Bulgaria

Analyzing the table regarding accessibility indicators for segregation in Roma ghettoized structures in settlements of Northwestern Bulgaria, we can derive insights into how accessible crucial services are within these communities, see Table 5. The services considered are emergency services, pre-hospital care, schools, and kindergartens, with a ranking system where a score of 3 indicates the worst condition (least accessibility) and a score of 1 indicates the best condition (most accessibility).

The towns of Belogradchik, Krivodol, Lom (Humata), and Miziya have the highest level of inaccessibility (score of 3) for all four services. In another group we observe Byala Slatina, Berkovitsa, see Fig. 8, and Varshets that show somewhat better access to schools (score of 1) but still face severe challenges with other services. Neighborhoods in Kozloduy and Montana (Ogosta) exhibit slightly better conditions for emergency services and pre-hospital care (score of 2) compared to other areas as they have better accessibility to the hospitals.

Kindergarten accessibility is notably poor across nearly all locations, with most areas scoring a 3. School accessibility shows the most variability, with some neighborhoods scoring better (1), indicating relatively better access to educational facilities. The "Mladenovo" neighborhood in Lom has the best access to kindergarten compared to all Roma GUS.

There are other cases where accessibility to school is not so easy. In the town of Byala Slatina, Vratsa district where the closest kindergarten is 2 km away from the neighborhood, see Fig. 9.

About 3000 Roma live in the town. They are increasing in recent years, but there is also a trend of emigration abroad—Italy, Spain, and Germany. There is a Roma

Table 5 Scores of accessibility indicators for segregation by neighborhood

Roma GUS	Emergency	Pre-hospital care	School	Kindergarten
Belogradchik	3	3	3	3
Byala Slatina	3	3	1	3
Krivodol	3	3	3	3
Oryahovo	3	3	2	2
Berkovitsa	3	3	1	3
Varshets	3	3	1	3
Kozloduy	2	2	1	3
Vidin	3	3	2	3
Dunavtsi	2	2	2	3
Lom (Mladenovo)	3	3	1	1
Lom (Humata)	3	3	3	3
Lom (Momin brod)	3	3	3	3
Lom (Stadiona)	3	3	2	3
Montana (Kosharnik)	3	3	2	3
Montana (Ogosta)	2	2	2	2
Miziya	3	3	3	3

1—GUS area is above 50% of proximity standards to specific site
2—GUS area is partially (below 50%) of the standards
3—GUS area is completely outside the normative standards

Fig. 8 GUS of Berkovitsa—neighborhood "Rakovitsa" the main street and a primary dchool "Georgi Rakovski"

Fig. 9 GUS in Byala Slatina

neighborhood that was established in the 1970s. The houses are legal, each house has a lot of electricity. The area is mostly covered by asphalt, and a sewerage system is currently being built. About 90% of the Roma live in the GUS. Religiously, there are both Christians and Muslims—something typical for Vratsa district area. They do not attend a segregated school. Adult Roma are mainly employed in construction as general workers, others in agriculture (in tobacco processing) or as municipal employees in the garbage collection department. Some of them are international drivers.

Nearly all Roma GUS exhibit poor accessibility to emergency and pre-hospital care, which can have dire consequences in health emergencies. Another observation is while some neighborhoods have better access to schools, the overall access to early childhood education (kindergartens) remains a significant issue, potentially impacting early childhood development and long-term educational outcomes. Potential focus areas for intervention can improve access to kindergartens and healthcare services and should be a priority for policymakers to address disparities and enhance the quality of life within these communities.

Conclusion

In the post-socialist countries, one of the main problem related to the Roma population is the accurate delineation of Roma neighborhoods, the identification of the number of inhabitants and the establishment of the actual housing situation and access to various urban resources and services. The results of the approach applied in this chapter are potentially useful for local authorities, urban planners, geographers, and architects responsible for urban planning and design. The mapping of ghettoized structures highlights the potential of effectively monitoring the multidimensionality of this urban phenomenon through the use of remote sensing and other data sources, given that information on ghettoized structures and their dynamics is often scarce or missing. The more rapid growth of the Roma population is leading to unequal provision of both housing and the provision of and access to various services, leading to a further increase in inequalities in society.

In Northwestern Bulgaria live different ethnic and religious communities, each with their own peculiarities and traditions. One of these ethnic communities is the Roma. In Europe, their number is estimated at around 10–12 million, with about 5% of them in Bulgaria. The problems associated with the Roma have many facets and their solution is very complex. The situation in Bulgaria is complicated by the fact that it ranks among the countries with the highest absolute number and relative proportion of Roma, on the one hand, and, on the other, their aggravated socio-economic situation inherited from the past.

A major problem in assessing the situation of the Roma ethnic group is the lack of data. One of the reasons for this lies in the peculiarities of the Roma ethnicity, which is not a homogeneous ethnic group. The Roma are a specific community that has no analogue with other ethnic groups in Bulgaria and in Europe. They are divided into many different and sometimes even opposing groups, subgroups and metagroups with their own ethnic and cultural make-up. Historical, political, and economic factors have created great differences among Roma. Historical, political, and economic factors have contributed to significant differences among Roma communities, further amplified by variations in language and culture. These disparities make it challenging to identify members of the Roma ethnic group, as they are characterized by weak ethnic consolidation processes and a limited sense of shared identity. However, the general attitude of society toward Roma, expressed in discrimination, reinforces their feeling of belonging to the same community. It would hardly be wrong to say that the unifying element in the Roma ethnic group is the macrosocial attitude toward it, thus forming a distinct social category that has common problems to solve. These are related to poverty, marginalization, exclusion from the labor market, segregation in the education system, difficult access to social services, extreme forms of spatial segregation, exclusion from the right to exercise active citizenship, intolerance, discrimination, prejudice, negative stereotypes, etc.

Until now, no mapping has ever been done in Roma neighborhoods in Bulgaria. This is a serious problem for the country and other European countries. This chapter proposes an approach to assess the current situation at three levels—object, neighborhood, and environ. Indicators for spatial segregation only are calculated. On the basis of 27 primary and secondary indicators, the spatial segregation of 16 ghettoized urban structures in three districts of Northwestern Bulgaria is calculated. The methodology used can be applied to other GGS, especially in post-socialist state structures. The largest segregation index is observed in the "Kosharnik" neighborhood in Montana and "Iztok" in Varshets while the lowest is calculated for Kozloduy and "Mladenovo" in Lom.

It is also important to note that while slum research continues, most studies focus only on a particular area in a particular city. Few have examined multiple slums in multiple cities (as in this article), and even fewer have examined how these territories change and evolve over time. The approach developed in this chapter can be used as part of a broader framework for understanding slums in this regard. For example, the indicators used to characterize slums can be examined over time and used to assess the impact of policies or other interventions in slums.

Acknowledgements This study has been carried out with a project titled "Spatial Models of the Roma Ghettoized Urban Structures in Bulgaria", funded by the Bulgarian National Science Fund under Contract No. KP06-H65/7 from 12 December 2022.

References

Asenov K (2018) Anthropology of the Ghetto—space and culture. Plovdiv: Studio 18 (in Bulgarian). 348

Bardarov G, Tsvetkov K (2020) Policies and measures for Roma integration—the European experience. In: Smart geography, vol: 100 years of the Bulgarian geographical society. Springer, pp 193–200

Bourdieu P (1989) Social space and symbolic power. Sociol Theory 7(1):14–25. https://doi.org/10.2307/202060

Castells M (1972) La question urbaine. Maspéro, Paris, Sociologie Du Travail Année 14–4:471–474

Durkheim E, Mauss M, Needham R (1963) Primitive classification. University of Chicago Pr., Chicago, p 143

Francini MCh (2013) Segregation policies in Sweden & Italy Supervisor: Prof. Jan Nilsson. Master thesis, Blekinge Institute of Technology, 40. https://www.diva-portal.org/smash/get/diva2:832901/FULLTEXT01.pdf. Accessed 3 April 2024

Harvey D (1973) Social justice and the city. Arnold, London, p 335

Harvey D (1982) The limits to capital. Blackwell, Oxford, p 478

Haussermann H (2005) The end of the European city? Eur Rev 13(2):237–249

https://kais.cadastre.bg/en

https://nsi.bg/sites/default/files/files/pressreleases/Census2021_ethnos.pdf

https://unhabitat.org/sites/default/files/2020/01/habitatiii-regional-report-europe-region.pdf

https://www.nsi.bg/bg

Ilieva N, Bardarov G (2020) (Ne)vazmozhnata romska integraciya: Demografski spetsifiki. Freiedrich Ebert Stiftung, Sofia. https://bulgaria.fes.de/fileadmin/user_upload/documents/publications/Nevazmozhnata_romska_integracia_Web_BG.pdf (in Bulgarian). Accessed 4 April 2024

Ilieva N, Asenov K, Kazakov B, Lubenov T (2020) Dynamics of the internal structure and habitation conditions of the Ghettoized urban structure of Harman Mahala quarter—Plovdiv, Bulgaria (through combined application of remote sensing and field research methods). In: Smart geography, vol: 100 years of the Bulgarian geographical society. Springer, pp 175–193

Knox P, Pinch S (2010) Urban social geography: an introduction (6th ed.). Routledge, New York and London, p 392

Kohli D, Sliuzas R, Kerle R, Stein A (2012) An ontology of slums for image-based classification. Comput Environ Urban Syst 36:154–163

Marcuse P (2005) Enclaves yes, Ghettos no: segregation and the state. In: Varady DP (ed) Desegregating the city: Ghettos, enclaves, and inequality. State University of New York Press, New York, pp 15–30

Massey DB (2005) For space. Sage, London, p 232

Massey DS, Denton NA (1993) American apartheid: segregation and the making of the underclass. Harvard University, Cambridge, MA, p 292

Peach C (1975) Urban social segregation. Longman, London, p 444

Stoyanova P (2017) The Roma people in the years of socialism: the policy of the Bulgarian state towards the Gypsy minority (1944–1989). Sofia, Paradigma, pp 255–258 (in Bulgarian)

United Nations (2017) Housing and urban development in the economic commission for Europe region. Europe, North America and Commonwealth of independent states towards a city-focused, people-centred and integrated approach to the new urban agenda, habitat III secretariat

Queer Lives: Navigating Identity, Acceptance, and Challenges in Rural Post-socialist Slovenia

Peter Kumer and Shé Mackenzie Hawke

Abstract Because 'Geography … attends to the multiple intersections and interactions within and beyond place' (Preston in Geogr Educ 28:41–49, 2015), we see it as a pivotal connective discipline in examining challenges, opportunities, and complexities in particular communities that raise specific concerns. With increasing movement between geographies of place and belonging, the possibilities of estrangement and alienation from one home base emerge, so that place becomes a contested site. Within the context of mobility between urban and rural places in Slovenia, the chapter addresses the recent post COVID-19 turn to 'settle' through the lens of rural in-migration. The chapter offers new conditions of possibility to increase inclusivity in educational settings that teach both traditional geography and "alternative geographies," as well as within community life, by both critically analyzing current trends and by interviewing diverse subjectivities in rural post-socialist Slovenia to develop data sets that can inform future curriculum design. Our study challenges the dominant narrative of rural areas as inherently less safe or welcoming for queer individuals by unveiling the nuanced realities of rural gay men in post-socialist Slovenia. Highlighting the importance of local contexts and the evolving nature of rural communities, this research contributes to a deeper understanding of the dynamics shaping queer rural lives and underscores the need for inclusive approaches in both academic and community settings.

Keywords LGBTQI+ · Rural subjectivities · Sexuality and space · Metronormativity · Rural in-migration

P. Kumer (✉)
Department of Geography, Faculty of Arts, University of Maribor, Maribor, Slovenia
e-mail: peter.kumer@um.si

S. M. Hawke
Department of Gender and Cultural Studies, University of Sydney, Sydney, Australia
e-mail: she.hawke@sydney.edu.au

© The Author(s), under exclusive license to Springer Nature Switzerland AG 2025
D. Piróg and A. Świętek (eds.), *Future-Ready Geography*, Key Challenges in Geography, https://doi.org/10.1007/978-3-031-81268-2_12

Introduction

Geography in recent times has expanded, much like biosocial anthropology, (in which social anthropological history and geographical fields merge as a single field of study) (Ingold and Palsson 2013), to include sub-fields such as cultural geography (Bissell and Gorman-Murray 2019; Massey 1994). It is to the notion of cultural geography that our chapter speaks and hopes to enrich by examining the coming together of particular 'places' and 'peoples'. Building social and cultural resilience in place-based education (Orr 1992) and geographies involves an extension of pre-existent and traditional geographical and cultural frames of reference. Because 'Geography ... attends to the multiple intersections and interactions within and beyond place' (Preston 2015), we see it as a pivotal connective discipline in examining challenges, opportunities, and complexities in particular communities that raise specific concerns. With increasing mobility between geographies of place and belonging, the possibilities of estrangement and alienation from one home base emerge—place becomes a contested site. Within the context of mobility between urban and rural places in Slovenia, the chapter addresses the recent turn to 'settle' in one place to resolve issues of belonging. We discuss this through the lens of rural in-migration (movement of individuals from urban to rural areas).

The recent COVID-19 pandemic saw an increase in 'one place' homing partly due to the restrictions of travel during the pandemic, and partly possibly because people had time to reflect on what made them happy. With an increase in social, geographic, and cultural flux impelled by climate change, local and global conflicts, food security issues and biodiversity loss, increasing transnationalism, and the recent COVID-19 pandemic, future-proofing coherent societal structures seem obvious. Yet, the coming together of these concerns in scholarly research and practice seems underrepresented. We make our start in this chapter by addressing these concerns, not so much as tensions, but as opportunities for resilience building and adaptation and providing a more diverse future for the studies of geography both in schools and at the tertiary level as well as through everyday life. We begin this enterprise by focusing on the subculture of LGBTQI+ communities in rural and urban Slovenia post the COVID-19 pandemic.

Geographic discussions on queer temporalities have predominantly explored transitions from rural to urban settings, highlighting urban areas as perceived havens of tolerance contrasted with the often heteronormative, suspicious, and secretive nature of rural spaces for queer individuals (Halberstam 2005; Weston 1995; Kazyak 2011). This urban preference, described by Halberstam (2005) as "metronormativity," stems from rural areas being viewed by some researchers as environments of persecution for queer people (Annes and Redlin 2012; Bell and Valentine 1995; Gorman-Murray 2007; Gorman-Murray et al. 2008; Kramer 1995).

In urban settings, there often appears to be a greater acceptance of diverse sexual orientations and gender identities, which can provide a sense of belonging and safety for queer individuals (Halberstam 2005; Hawke and Offord 2012). These spaces can offer a contrast to the heteronormative structures of rural areas, and thus, the act of migrating to urban areas itself can be significant life event that marks a form of 'coming out' or finding community.

As highlighted by Gorman-Murray (2007), this broad generalization largely arises from studies carried out in highly urbanized countries, such as USA and the United Kingdom.

Mostly western studies on same-sex desire in rural settings have examined the realities faced by rural queer individuals, highlighting challenges such as scarce access to queer communities, stigmatization and homophobic atmospheres, along with incidents of physical and verbal abuse, and the concealment of one's homosexual identity out of fear (Rickard and Yancey 2018). Yet this is not always the case, as LGBTQI+ scholarship from the Northern Rivers region in Bundjalung Country, NSW of Australia has demonstrated (Hawke and Offord 2012). The Northern Rivers is an exceptional example of rural in-migration and queer and inclusive belonging and constitutes one of the largest populations of LGBTQI+ people outside of the major Australian cities and indeed global cities.

This notion that LGBTQI+ individuals residing in non-urban regions are at disadvantage compared to their urban counterparts has been contested in several studies, uncovering that many LGBTQI+ people who live outside the big city are satisfied with their daily lives (e.g. Champion and Hugo 2016; Gray et al. 2016). Over time, rural locales are lesser often seen as hostile to queer communities, indicating that openly homosexual identities can be accepted and lived out in specific rural environments. Klotz (2021) interprets the perceived shortcomings of rural areas—such as their slow-paced, lack of modernity and trendiness—as a chance for rural queer individuals to break free from the restrictive constraints of metronormativity, that is, assuming there are greater benefits by living in a city (Halberstam 2005). Essentially, the unique pace and nature of rural life foster and support the existence of queer identities.

However, Kuhar and Švab (2014) have noted that the perceived dichotomy between rural and urban areas as it relates to the queer experience is predominantly a Western concept and may not fully translate to other settings. In Eastern European countries, there are instances where queer people perceive rural regions as being more accepting and safer than urban ones where queer individuals often find urban spaces less welcoming due to less visible gay culture in cities and higher levels of urban homophobia and violence. For example, despite the overall increase in tolerance toward different sexual orientations in Slovenia in recent years, the pride parades in two largest cities in 2023 were marked by violence: storefronts were shattered, rainbow flags were burned, participants were subjected to physical and verbal attacks, and eggs were thrown (STA 2023).

One aspect not fully addressed yet is the potential for a fulfilling life in rural areas for queers who choose not to migrate to large cities and do not adhere to the standards of Western urban "gay culture." Hence, the setting for our chapter and the challenges and potentials it present is rural post-socialist Slovenia. We explore the subjectivities of queer (primarily gay males) people who live in the rural areas and small towns in Slovenia three decades after the dissolution of Yugoslavia, which started in 1991.

This period has opened opportunities for democracy and human rights, free from the constraints of the Yugoslav era, and is subsequently more inclusive.

The chapter draws on in-person interviews with gay men, to explore their understanding and navigation of daily life within their particular rural or small-town settings. We acknowledge at the outset the challenges faced in recruiting women, bisexual, and other individuals with LGBTQI+ identity for this research—a point further elaborated in the methods section. This is not to exclude women, rather it is to focus on men in this chapter that we aim to follow on with a further chapter on gay and lesbian women and subjects that are included in the LGBTQI+ identity umbrella. We focus on the following thematic aspects: (1) the reasons queer individuals opt for rural life despite its inherent challenges; (2) the negotiation of queer presence within rural settings; and (3) the importance of city life to queer identity. Besides the thematic aspects, the aim is also to explore routes to achieve qualitative richness in human geography with populations that are "hard-to-reach," necessitating the creation of a "safe" environment for sharing deeply personal information.

Perspectives on Quality of Rural Life in Central and Eastern Europe

Our analysis builds on approaches that have demonstrated how different socio-cultural, political and economic factors have played an integral role in the shaping of a traditional rural landscapes that arguably still exist in Central and Eastern (CEE) European countries (Palang et al. 2006).

Rural areas in CEE differ from those in Western Europe primarily due to their historical trajectories and socio-economic changes. These countries have experienced rapid changes in the twentieth century, (not the least of which is the consequences of multiple wars) leading to unique landscape characteristics that include a mix of traditional rural landscapes alongside signs of industrialization and urbanization and population changes as a result of war and displacement. This contrasts with Western Europe, where industrial development and urban sprawl have more uniformly transformed rural landscapes (Palang et al. 2006). The abundant natural areas and rural populations contribute to the preservation of rural identities.

Although post-socialist CEE countries have not reached the growth of the urban population as much as the western capitalist societies, Slovenia stands out even among CEE countries in being among the least urbanized countries measured either through high population density per km^2 or through percentage of population living in an urban area (Potočnik 2001; Uršič 2012). According to Eurostat (2016), Slovenia is among the most rural countries in EU. Although urbanization in Slovenia accelerated following the Second World War and especially in the 1960s and 1970s due to socialist modernization which brought relatively rapid (though late in European context) industrialization (Klemenčič 1991), the urban system never achieved a level of high urbanization (Dijkstra and Poelman 2014). To illustrate, today there is more than 90% of settlements out of 6035 with 500 or less inhabitants while on the other hand only two cities (Ljubljana and Maribor) comprise more than 50,000 inhabitants (Miklavčič 2016).

The factors influencing a high level of rurality have been extensively examined in research. The key factor according to Klemenčič (2006) is planned polycentric development after the Second World War with spatially dispersed industrialization. This resulted in fragmented land ownership and reduced agricultural land, decreasing the number of pure farming households and significantly reducing the population dependent on agriculture. It is broadly recognized that daily commuters transformed the countryside. With one of the EU's highest rates of car dependence ("automobilization," Uršič 2012) and among the lowest uses of public transit (Bole and Gabrovec 2014), Slovenia also ranks very high in the EU for road investment per person (Rudolph et al. 2023). The rapid expansion of the highway network in the 1990s contributed to widespread car usage, encouraging the shift from urban to rural living (Uršič 2012).

Living in single-family homes in small, remote settlements is proven to be a preference within the Slovenian social value system (this was termed "anti-urbanism" by Uršič and Hočevar 2007). Being close to nature is seen as the most desirable aspect of where to live (Plostajner et al. 2019). The convenience of maintaining employment in urban centers, which are accessible from rural areas within an hour's drive, is supported by legal financial incentives for commuting (Bole and Gabrovec 2014). This encourages moving to the outskirts as it can be financially beneficial when buying a home. Yet, concerns have been raised about the sustainability of such spatial development patterns. Dependence on cars and the high costs of providing infrastructure and services to small, distant communities are seen as neither environmentally nor economically sustainable (Plostajner et al. 2019).

In Slovenia, high levels of rurality are also partly influenced by perceived threats, with concerns like unemployment and poverty being more prevalent among urban residents than those living in the rural areas (Bučar Ručman 2019). The seasonal beauty of the countryside is also a strong geographical drawcard.

Another important factor that the push toward urbanization was less intense than in other socialist and Western European nations is partly because there was no layered social structure, including an aristocracy and bourgeoisie (urban fabric). This situation fostered a deep sense of connection with rural areas (Uršič 2015) and supported the development of values that are somewhat resistant to urban influences and favor equality (Szelenyi and Kostello 1996; Uršič and Hočevar 2007).

Queering Rural Slovenia Within the Post-socialist Perspective

Slovenia led the way in socialist Yugoslavia by decriminalizing homosexuality in 1976, well before the country dissolved in the 1990s. It also launched the first gay and lesbian movement in socialist Central and Eastern Europe in 1984, marked by a festival of international gay-themed movies that continues today (Lešnik 2006; Kuhar 2016). This progressiveness is linked to Slovenia's economic prosperity and its strong ties with the West, notably Italy, post-1975 Osimo Treaty, setting it apart from other socialist states with its relatively open borders.

Slovenia's capital held the first large-scale Pride Parade in the former Yugoslavia in 2001. Despite two failed referendums on same-sex rights in 2012 and 2015, Slovenia legalized same-sex marriage in 2022 following a decision by the Constitutional Court, marking a first in the former Yugoslav nations.

Nearly four decades after the inception of its gay and lesbian movement, Slovenia still has limited "gay infrastructure," as described by Kuhar and Švab (2014), especially when compared to Western nations. Such amenities, including gay bars, saunas, as well as LGBTQI+ non-profits, are found primarily in Ljubljana (with an exception of event nights and activities occasionally extending outside Ljubljana). Slovenia's second-largest city, Maribor, along with other areas, lacks these facilities almost entirely.

Slovenia's rural character and Ljubljana's status as a "provincial town" in a European context have according to Kuhar and Švab (2014) led some gay and lesbian individuals to seek opportunities abroad, at least temporarily. This trend reflects a broader historical perspective where Western Europe was seen as offering a "better life" compared to the "symbolically rural" homeland, a sentiment rooted in the socialist era. This perception underscores the ongoing urban-rural divide in the country's LGBTQI+ community dynamics.

Studies reveal a distinct rural-urban voting pattern in Slovenia, with rural areas more likely to support conservative political parties, according to Uršič and Tiran (2022). Acceptance of LGBTQI+ individuals also varies, with rural areas typically showing less tolerance and more verbal hostility. However, Kuhar and Švab (2014) note that these disparities are gradually diminishing.

Methods

In this study, we adopt an "interpretivist approach," prioritizing understanding reality through individuals' lived experiences. This method acknowledges that people interpret the same situations differently, highlighting the need for in-depth exploration of these diverse perspectives to understand human experience complexity. To achieve "rigor," we utilized a semi-structured interview format, allowing customization to each participant's context, (or their ethnographic field) ensuring detailed, and what

Geertz calls "densely textured facts" and "thick descriptions" (Geertz 1973 p. 76) of cultural meanings and behaviors.

We examined the stories gathered through a geo-ethnographic lens (Crang 2007) to deeply explore how people interact with their rural environments, highlighting the significant role of geography, culture, and social frameworks. This approach helps in understanding the intricate dynamics of rural life beyond what quantitative research might reveal, acknowledging the rich, contextual interconnections that shape local experiences.

Notes on emerging ideas were kept throughout to enhance the analysis's depth and reliability.

Participant Recruitment and Establishment of a »Safe« Interview Environment

Gaining access to hidden population that this chapter is addressing (Švab and Kuhar 2008) and then ensuring they felt safe during interviews, allowing them to express themselves freely, has been among the most significant challenges encountered in this study.

Finding rural queer participants (in this instance gay males) is challenging and potentially ethically complex (Gorman-Murray et al. 2013). This is why studies on queer lives, even for understanding lives in rural areas, often involve participants from urban settings (see for example Butterfield 2018 and Marlin et al. 2023). This reflects a significant gap in research that our study aims to address.

Our research encountered similar challenges. While identifying queer participants in urban areas was relatively straightforward, we specifically sought rural individuals to meet our criteria, leading us to exclude urban candidates. Despite approaching several rural queer individuals, they declined to participate. This hesitancy may stem from concerns about maintaining anonymity, as achieving privacy might appear more challenging in smaller, rural communities than in larger, urban populations where blending in is easier. On the other hand, two interviewees did not ask for anonymity as they were "out" to local community. The difficulty to reach them stems from their hard availability due to work on their farms.

A significant obstacle was the difficulty in engaging representatives from the full LGBTQI+ community. This effort was abandoned when several LBT+ individuals declined to participate. For this purpose, while we still employ the umbrella term queer, we are aware of the limits of participation in this particular research chapter. Consequently, we adopted an alternative strategy. The first author of this chapter utilized their personal social networks and lived experience, which resulted in the recruitment of only rural gay men, although none of the interviewees had a prior acquaintance with the interviewer. Therefore, the type of sampling we employed, is purposive (Weiss 1995). Despite significant challenges in finding suitable participants, we managed to conduct five interviews. This number was deemed sufficient for reaching saturation, as defined by Braun and Clarke (2021), where no additional novel insights were observed.

Considering the significant role of the researcher's positionality in ethnographic studies (Rose 1997), especially when addressing sensitive issues like sexuality, we adopted an "empathic approach" (Miles 2020). We implemented this strategy while strictly adhering to professional boundaries and complying with institutional ethical guidelines. This fosters an environment where participants felt secure and understood, thereby encouraging them to divulge personal experiences and sensitive data with a trusted person from their community.

Those who volunteered to be interviewed were assured of complete anonymity. We proposed in-person meetings but also gave them the option to choose a time and location where they felt most at comfortable. Two of the interviewees opted for interviews within their homes, another welcomed us into their workplace for the interview, one preferred video conferencing technology, and another chose to come to our office. In contrast to the findings by Miles (2020), individuals who were not openly "out" or were selective about being "out" all opted against meeting in public spaces.

While offering incentives (institutional promotional items valued at 20–30 euros) aimed to boost participation, it was the interviewer's university affiliation that truly built trust. Being linked to an academic institution naturally instilled confidence in the ethical conduct of the research, making participants more willing to share intimate stories. This trust was echoed by a volunteer who mentioned before the interview their willingness to assist researchers without needing rewards, highlighting the importance of research for societal progress.

To safeguard the privacy of our interviewees, we will avoid mentioning the name of the village where they reside, since revealing this could potentially allow people familiar with these small villages to identify them, rather than protect their requested privacy. In some cases, we only use the name of the regions where the villages are located. We will also refrain from revealing the real names of interviewees as this could also jeopardize their anonymity. Nonetheless, some details can be shared about the participants; they are all of Slovenian nationality or descent, originating from western Slovenia or areas bordering Slovenia, and they currently reside in rural regions. While two interviewees live in the rural areas since their birth, the three moved to the rural areas from what they consider larger urban area. Aside from one participant living in the suburbs, the rest are situated in areas typically referred to as "peripheral" (Klemenčič 1991; Klemenčič et al. 2008), contrasting with "central" areas. Three of our interviewees are involved in agriculture, either full-time or part-time, and two of our interviewees work in non-agricultural sectors. Most of them commute to work, at least a few days a week, to bigger cities, which are by far the largest centers in the region (population 100,000, 200,000 and 300,000), whereas another commutes between two villages: "*from a village with 18 people to one with just four inhabitants.*" The participants are from varied age groups: one is in his thirties, two are in their early forties, one is in his early fifties, and one is in his late fifties.

Interview Process

Sets of questions were designed to gain deeper insights into interviewees' connections with the rural land and their experiences living there. These questions prompted participants to share their life stories, providing detailed insights into personal experiences, as well as individual perceptions (Connell and Messerschmidt 2005). Given the frequent occurrence of migrations from rural to urban areas and internationally in the "queer quest for identity" (Knopp 2004), we utilized the "life course approach" (Wimark et al. 2017; Lewis 2017; Paniagua 2023) to explore the role of urban environments in the lives of rural gay men. "Integrated" life course approach was considered where fieldwork context, experiences, and the life course of both the researcher and the subjects influence each other reciprocally.

Several questions aimed to elicit descriptions of the contrasts between cities and rural areas, and, considering the interviewees are part of a "rural other," how perceived acceptance might vary. Other inquiries delved into their interactions with the wider rural community which yielded a spectrum of responses, not just binarized "otherness." Interviews lasted between 30 and 45 min.

Data Analysis

We organized interviews and field notes digitally. Interviews were audio recorded and transcribed non-verbatim. Quotes have been modified for confidentiality and clarity without altering their original meaning or content. To analyze qualitative data, we used ATLAS.ti software. Coding included a two-step process. First, the Open Coding process included reading each interview line-by-line to maximize the extraction of outcomes. This step helped identify new and potentially previously unknown perspectives of rural queer lives. The second step included the process of reorganization, merging and grouping of the codes, forming a hierarchy of codes that grouped data by similar themes. We then aligned our findings with existing literature to better grasp the rural queer community's experiences and viewpoints.

Findings

Why Do Gay Male Individuals Choose Rural Life Despite Its Challenges?

Acknowledging the idea that living in rural areas poses challenges for "rural others" such as queer individuals, it is interesting to explore people's reasons for choosing such a lifestyle. One participant, in his early forties, moved from a Slovenian town of medium size (with a population of 5–6000) to the countryside due to a series of events. His partner was born in the area, worked there, and owned a house with a small farm. Despite considering various options, including living in a large city (population 200,000), their preference was to remain close to nature and continue farming.

> We both appreciated the opportunity to be close to nature, to enjoy peace outside our home without the hustle and bustle of a large city, and to have more than just jobs and a living space but also a deeper connection to nature. We already had animals and harbored a desire to make something out of them, not just as an expense but to create and work with them.

The desire for a quiet life in nature, far from the hustle and noise of the city, was a key factor for moving to the countryside for other respondents as well. One participant, in his late fifties, who had lived in a city until he was 45, purchased a house in a small village with only 50 inhabitants just two months after his initial visit:

> I came here because I liked the environment where I live. Here's a lot of forest, greenery, and open, undeveloped land. It was beautiful and peaceful. Trieste is chaotic. I couldn't stand it anymore. Now, when I go down to Trieste, I leave as soon as I can.

The interviewee, a man in his early forties from an ethnic minority that fought to preserve its identity against assimilation by a larger nation, values his territorial identity more than his sexual orientation, describing it as just one aspect of himself but not the defining one. He expresses a deep connection to his village, seeing it as a place where are:

> …my roots and my family. I like the village because the house where I was born is there. Because it's my land. Sometimes, I've also wondered if this is somehow psychologically linked to the idea that gays might have a harder time having descendants. So, what remains is the land. It's the cradle where you were as a baby.

This expression reveals a profound connection to one's origins and the contemplation of legacy and belonging within the context of personal and communal identity.

Interestingly, our interviewees are not worried about the potential negative impacts of rural areas' heteronormative attitudes on their wellbeing. They shared that they have never encountered negative reactions in villages; in fact, their experiences were quite the opposite, as one interviewee noted:

> Everyone knows [that they are a couple], and we're met with positive attitudes, which is reassuring. At times, we might be overly influenced by a fear we've internalized about potential reactions, and we may even misinterpret some responses. Around here, as we walk about or go anywhere, everyone greets us as though we're acquainted, even though we don't actually know them [laughs]. This could be seen in a negative light.

How is Gay Male Presence Negotiated in Rural Environments?

While numerous studies from the "Western" perspective have identified a shift toward a 'post-gay' and 'beyond the closet' era (Seidman 2002; Ghaziani 2014), highlighting a normalization aligning with neoliberal values that promote uniformity and lessen the radical demands of LGBTQI+ movements (Richardson 2005), our research focuses on exploring the presence and characteristics of this normalization in rural areas within a post-socialist context.

To understand if rural geographies of "neglected others" can still be conceptualized as "the closet" (Cloke and Little 1997), involving concealment, disconnection from other gay individuals, internalized shame, guilt, fear due to societal attitudes toward homosexuality, and leading a double life, we investigated further. The Periodical Slovenian Public Opinion Survey has found that in the nineties, more than 60% of people did not want to have a homosexual as a neighbor; now, this proportion has dropped to just over 20% (Hafner-Finkc 2022). Yet the respondent, in his early forties, lives in a community that he describes as very conservative, attributing this to their lifestyle:

> In the area I come from, farmers traditionally spoke my language and worked the land. This definitely contributed to our culture. Regarding Christianity, during extreme nationalisms, priests protected us, ensuring we could learn our faith in Slovenian when outside it was often not allowed, or you'd face ridicule. Priests defended us, making the church a refuge. I still sing in the church choir and attend mass regularly, as we were raised in this tradition.

Just one participant avoids talking about his sexuality with people in their village entirely, even when directly asked. All other respondents, however, indicated that while they do not feel the need to publicly declare their sexuality, they also do not engage in behaviors that would make them appear straight or conform to heterosexual norms simply for the sake of blending in. As one respondent in their late 50s shared:

> In the village, they know. When I mentioned going to the city, they knew why. I didn't specify anything; I don't talk about my life. If asked, I'll tell, but I don't openly announce that I'm gay. I don't agree with that approach. Now, I think all 50 people know I go to the city because of my boyfriend. But I never brought it up.

Rural queer individuals being subtly open about their identity, presenting themselves as "*just another local,*" serves as a strategy to foster deeper community ties, which Gray (2009) identifies as the "rural politics of recognition" for community acceptance. This contrasts with the "out and proud" attitude that stresses uniqueness to gain acceptance. An interviewee pointed out that

> …people [Heterosexuals] get annoyed because we [LGBTQI+ individuals] want to stand out, like with the pride parade. Straight people don't have their own parades. If you leave others alone, they'll leave you alone.

Yet, it prompts us to consider if seeking recognition in rural areas involves gaining acceptance by downplaying non-traditional identities. This is particularly relevant in a post-socialist context, where for example pride parades, unlike their more commercial and festive counterparts in the West, still embody political protest. In Ljubljana, for instance, these events maintain their political significance and form, contrasting with the depoliticized nature of similar events in Western countries.

Is City Life Essential for Gay Male Identity?

To challenge the notion that queer visibility is more accepted in urban areas, where being "out" is often easier for queer individuals, we sought to examine the rural queers' relationships with large cities. One part that we were particularly interested is the perception of our interviewees of larger urban areas.

Our respondents do not see the cities as places that would suit their nature. The respondent in his early forties highlighted his preference for a way of life grounded in heritage and community values over the urban environment, regardless of the city's size or significance. He emphasized his deeper connection to a community influenced by their territorial identity:

> Comparing city life to where I work, even though it's not a major city like Milan, Paris, or Vienna, I don't feel a strong connection to that urban lifestyle. Coming from a background deeply rooted in traditions, old habits, and being part of the Slovenian minority group likely influences this sentiment.

The same sentiment was identified with other interviewees, for example an interviewee in his early forties who was together with his boyfriend for a while torn between choosing the [rural areas] or [big city], but ultimately, both of them preferred

> ...being in a place where nature is accessible, where you can enjoy peace in front of your house, away from the hustle of a big city with its traffic, and have more than just work, an apartment, and leisure activities, but also a deeper connection to nature.

The respondent in his early thirties continues that he always loved cities. As a teenager, he dreamed of escaping to [the city], not due to any problems at home but because of the allure of city life, like partying in Ljubljana on weekends. However,

> ...living there seems to require too many compromises compared to what's important to me. /.../ Big cities tend to be more career-focused, which is more of a Western mindset. In a big city, you can lose touch with yourself, whereas here, I feel I've found a balance between work, personal life, and self-improvement. /.../ Big cities have advantages for LGBTQI+ freedom, acceptance, and expression, but personally, I don't feel I'm missing out on that where I live.

For their holidays, interviewees do visit cities which are according to interviewee in his early forties *"usually spots where [they] settle for a few days, which only confirms [their] preference for the tranquillity of rural life. However, sometimes you do need a bit of city life to feel the energy of being around others.«*

The purpose of traveling to these cities is not primarily to visit gay-friendly destinations, as explained by another interviewee:

> We usually decide to visit cities for concerts, choosing the destination based on that. We might explore the local scene a bit, but attending a well-known gay pride event isn't our primary reason for traveling there.

Given the significance of cities for LGBTQI+ individuals' sexual identity, we delved further into this issue. Questions were related on their use of location-aware, real-time dating apps for LGBTQI+ people, because as explained by the interviewee in his fifties the experience of using these apps in cities and the village is different and *"It's hard to find [a date]. You need to take the time and prepare well.«* An interviewee in his fifties shared his strategy of using these apps by changing his location settings:

> I had two profiles before, one for Trieste and another for Koper. When my location showed [my village], I got no messages. I met you [looks at his partner] with my location on Koper. Koper, being more populous, offered more opportunities. I even traveled to Ljubljana twice to meet people from the app.

This strategy of setting his profiles to larger cities underscores the importance of urban settings in facilitating social connections for LGBTQI+ individuals, contrasting with his lack of "success" when his location was set to his village. Yet, the ability to connect virtually lessens the necessity of living in urban areas. The significance of easy access to these cities is underscored by their proximity, as noted, "*Trieste, Ljubljana, Rijeka—they're all about 40 min away,*" indicating that physical distance to urban centers remains a practical consideration (Fig. 1).

According to interviewee in his early forties, cities are more tolerant because there are more people, which allows individuals to feel more at ease to be themselves. In urban settings, like the city where he works, he hardly knows anyone in his apartment

Fig. 1 Despite the perception of living in a secluded, nature-surrounded rural village, major employment centers are actually just 20–40 min' drive away. This proximity significantly enhances the appeal of rural living of LGBTQI+ persons, traditionally attached to urban spaces. The photography showcases the village of Tinjan (population 150) in the foreground and the city of Trieste (population 200,000) in the background

building, leading to less judgment due to anonymity. In contrast, village life involves closer scrutiny, with neighbors watching and commenting on each other's activities:

> When you go shopping, people watch from their windows and the next day on the street, they might say to you, Did you go shopping? This doesn't happen in cities.

Although this interviewee is expressing that cities are perceived as more tolerant due to their larger populations, which allows individuals to feel more comfortable being themselves, he goes on saying that:

> staying in the hometown is the best way to promote tolerance. Individual behavior can lead to daily, incremental changes and everyone has the power to influence their surroundings positively through their actions. /.../ There were occasions when I encountered mean jokes or those bordering on discrimination. In such cases, I responded, sometimes with jokes aimed at those making inappropriate remarks, to get them to stop.

Discussion and Conclusion

In this research, we delved into the lived experiences of rural gay men in post-socialist Slovenia, challenging the prevailing Western-centric narrative that urban environments invariably offer safer and more accepting spaces for queer individuals. Our findings reveal a nuanced reality where, contrary to conventional beliefs, rural areas can also serve as spaces of acceptance and community through the recent phenomena of rural in-migration during and after COVID-19. This contrasts sharply with the experiences commonly reported in Western contexts, emphasizing the need to consider local cultural and historical nuances in the study of queer lives.

Our findings contribute significantly to the understanding of rural queer lives, particularly in highlighting how community acceptance can vary dramatically based on geographic regional and cultural contexts. For rural gay men in Slovenia, the dissolution of Yugoslavia and the subsequent socio-political changes have created a unique environment where, in some instances, rural settings provide a safer or more welcoming atmosphere than urban ones. This challenges the binary opposition of rural versus urban queer experiences and suggests a more complex interplay of factors influencing acceptance and identity.

The study's insights into the negotiation of queer identity within rural settings also open up new avenues for support services and policy-making aimed at improving the quality of life for queer individuals outside urban centers. It calls for a re-evaluation of strategies aimed at fostering true inclusivity and understanding, tailored to the specific needs and realities of rural communities.

The inherent challenges faced in recruiting a broader spectrum of participants, particularly women, bisexual, and transsexual individuals, emphasize the complexities and limitations of our study. This difficulty not only reflects the potential fear and stigma still present in rural areas but also highlights the importance of creating safer environments for all queer individuals to share their experiences. It signals an urgent need for further research that encompasses the diverse spectra of identities within the LGBTQI+ community, ensuring a more comprehensive understanding that transcends the male-centric narratives often predominant in research.

Looking forward, this research lays the groundwork for more inclusive studies that aim to capture the varied experiences of the LGBTQI+ population in post-socialist and rural contexts. By acknowledging the limitations of our current study and proposing directions for future research, we underscore the need for a more nuanced understanding of queer lives across different geographical and cultural landscapes.

In conclusion, this study not only contributes to the burgeoning field of rural queer studies but also challenges scholars and practitioners to rethink the complex dynamics of acceptance, identity, and community within varied social and geographic settings. By extending our focus beyond the urban-centric narratives that have dominated queer research, we can begin to uncover the rich tapestry of experiences that define queer lives globally, particularly in underrepresented regions like rural post-socialist Slovenia.

References

Annes A, Redlin M (2012) The careful balance of gender and sexuality: rural gay men, the heterosexual matrix, and "effeminophobia". J Homosex 59(2):256–288

Bell D, Valentine G (1995) Queer country: rural lesbian and gay lives. In: Bell D, Valentine G (eds) Mapping desire: geographies of sexualities. Routledge, London, pp 113–127

Bissell D, Gorman-Murray A (2019). Disoriented geographies: undoing relations, encountering limits. Trans Inst Br Geogr 44(4):707–720

Bole D, Gabrovec M (2014) Geography of daily mobilities in post-socialist European countries: evidence from Slovenia. Mobilities in socialist and post-socialist states: societies on the move. Palgrave Macmillan UK, London, pp 217–239

Braun V, Clarke V (2021) To saturate or not to saturate? Questioning data saturation as a useful concept for thematic analysis and sample-size rationales. Qual Res Sport, Exercise Health 13(2):201–216. https://doi.org/10.1080/2159676X.2019.1704846

Bučar Ručman A (2019) Social ties, solidarity and threat perception in rural and urban communities in Slovenia. Revija za kriminalistiko in kriminologijo/Ljubljana 70(5):409–421

Butterfield N (2018) Imagined rural/regional spaces: non-normative sexualities in small towns and rural communities in Croatia. J Homosex 65(13):1709–1733

Champion T, Hugo G (eds) (2016) New forms of urbanization: beyond the urban-rural dichotomy. Routledge, London

Cloke P, Little J (1997) Conclusion: marginality and rural others. In: Cloke P, Little J (eds) Contested countryside cultures. Otherness, marginalisation and rurality. Routledge, London, pp 272–285

Connell RW, Messerschmidt JW (2005) Hegemonic masculinity: rethinking the concept. Gender Soc 19(6):829–859

Crang MA (2007) Doing ethnographies. Sage, London

Dijkstra L, Poelman H (2014) Regional working paper 2014. A harmonised definition of cities and rural areas: the new degree of urbanisation. WP, 1, p 2014

Eurostat (2016) Urban Europe—statistics on cities, towns and suburbs—the urban paradox. Retrieved from https://ec.europa.eu/eurostat/statistics-explained/index.php?title=Urban_Europe_%E2%80%94_statistics_on_cities,_towns_and_suburbs_%E2%80%94_the_urban_paradox&oldid=302194

Geertz C (1973) Thick description: towards an interpretative theory of culture. Basic Books, New York

Ghaziani A (2014) There goes the gayborhood? Princeton University Press
Gorman-Murray A (2007) Rethinking queer migration through the body. Soc Cult Geogr 8(1):105–121
Gorman-Murray A, Waitt G, Gibson C (2008) A queer country? A case study of the politics of gay/lesbian belonging in an Australian country town. Aust Geogr 39(2):171–191
Gorman-Murray A, Pini B, Bryant L (2013) In "Geographies of ruralities and sexualities." In: Gorman-Murray A, Pini B, Bryant L (eds) Sexuality, rurality, and geography. Lexington Books, New York, pp 1–17
Gray M (2009) Out in the country: youth, media, and queer visibility in rural America. New York University Press, New York
Gray ML, Johnson CR, Gilley BJ (eds) (2016) Queering the countryside. New Frontiers in rural queer studies. New York University Press, New York
Hafner-Fink M (2022) Mednarodni družboslovni anketni program ISSP in problemi primerjalnega družboslovnega raziskovanja. In: Toš N (ed) Vrednote v prehodu VII: Slovenija v mednarodnih in medčasovnih primerjavah SJM–ISSP 1991–2012. Fakulteta za družbene vede, Ljubljana, pp 21–26
Halberstam J (2005) In a queer time and place, transgender bodies, sub-cultural lives, vol 3. New York University Press, New York
Hawke SM, Offord B (2012) Querying pedagogy and cultural studies: critical reflections on teaching sexuality. In: Scherer B, Ball M (eds) Queer paradigms, vol. 11: interrogating agendas. Lang Publishers, Oxford, pp 191–208
Ingold T, Palsson G (2013) Biosocial becomings: integrating social and biological anthropology. Cambridge University Press, Cambridge
Kazyak E (2011) Disrupting cultural selves: constructing gay and lesbian identities in rural locales. Qual Soc 34:561–581
Klemenčič V (1991) Tendence spreminjanja slovenskega podeželja. Geografski Vestnik 63:25–40
Klemenčič MM (2006) Teoretski pogled na razvojne strukture slovenskega podeželja. Dela 25:159–171
Klemenčič MM, Lampič B, Potočnik Slavič I (2008) Življenjska (ne)moč obrobnih podeželskih območij v Sloveniji. Filozofska fakulteta, Ljubljana
Klotz M (2021) Considering (queer) musicking through autoethnography. MUSICultures 48:9–28
Knopp L (2004) Ontologies of place, placelessness, and movement: queer quests for identity and their impacts on contemporary geographic thought. Gend, Place Cult 11(1):121–134
Kramer J (1995) Bachelor farmers and spinsters: gay and lesbian identities and communities in rural North Dakota. In: Bell D, Valentine G (eds) Mapping desire: geographies of sexualities. Routledge, London, pp 200–213
Kuhar R (2016) The heteronormative Panopticon and the transparent closet of the public space in Slovenia. In: De-centring western sexualities. Routledge, pp 149–166
Kuhar R, Švab A (2014) The only gay in the village? Everyday life of gays and lesbians in rural Slovenia. J Homosex 61(8):1091–1116
Lešnik B (2006) Melting the iron curtain: the beginnings of the LGBT movement in Slovenia. New social movements and sexuality, pp 86–96
Lewis NM (2017) Linked life courses in fieldwork: researcher participant and field. Area 49(4):394–401
Marlin L, Lewis C, McLaren S (2023) "Being able to be yourself": a qualitative exploration of how queer emerging adults experience a sense of belonging in rural Australia. J Homosex 70(14):3306–3327
Massey D (1994) General introduction. In: Massey D (ed) Space, place and gender. Polity Press, Cambridge, pp 1–16
Miklavčič N (2016) Poročilo o prostorskem razvoju. Ljubljana: Ministrstvo za okolje in prostor. Available at: https://www.gov.si/assets/ministrstva/MNVP/Publikacije/proocilo_o_prostorskem_razvoju.pdf. Accessed 10 Jan 2024

Miles S (2020) "I've never told anyone this before": co-constructing intimacy in sex and sexualities research. Area 52(1):73–80
Orr DW (1992) Ecological literacy: education and the transition to a postmodern world. State University of New York, Albany, NY
Palang H, Printsmann A, Gyuro EK, Urbanc M, Skowronek E, Woloszyn W (2006) The forgotten rural landscapes of Central and Eastern Europe. Landscape Ecol 21:347–357
Paniagua A (2023) Laboratory life: twenty years of experimental research with locals and new comers in remote rural areas. J Geogr Res 6(2):1–14
Plostajner K, Medved P, Cerar A, Simoneti M (2019) Is Ljubljana being eco-gentrified?: the case of sustainable urban development of the city centre. Sociologia urbana e rurale: XLI 119:117–133
Potočnik I (2001) Geographical aspect of rural areas: transformation in Slovenia. Bulletin De La Société Géographique De Liège 41:69–76
Preston L (2015) The place of place-based education in the Australian primary geography curriculum. Geogr Educ 28:41–49
Richardson D (2005) Desiring sameness? The rise of a neoliberal politics of normalisation. Antipode 37(3):515–535
Rickard A, Yancey CT (2018) Rural/non-rural differences in psychosocial risk factors among sexual minorities. J Gay Lesbian Soc Serv 30(2):154–171. https://doi.org/10.1080/10538720.2018.1444525
Rose G (1997) Situating knowledges: positionality, reflexivities and other tactics. Prog Hum Geogr 21(3):305–320
Scherer B, Ball M (eds) (2009) Queer paradigms Vol. 11: Interrogating agendas. Lang Publishers, Oxford
Seidman S (2002) Beyond the closet: the transformation of gay and lesbian life. Psychology Press
STA (2023, 17 Sept) Maribor Pride Parade marred by violence. STA. https://english.sta.si/3214602/maribor-pride-parade-marred-by-violence
Švab A, Kuhar R (2008) Researching hard-to-reach social groups: the example of gay and lesbian population in Slovenia. Društvena istraživanja: časopis za opća društvena pitanja 17(3(95)):529–552
Szelenyi I, Kostello E (1996) The market transition debate: toward a synthesis?. Am J Sociol 101(4):1082–1096
Uršič M (2012) Urbanisation history of Slovenia in the context of Eastern and South-Eastern European countries. EQPAM 1(1): 38–47
Uršič M (2015) Skriti pridih podeželja? Razrednost in »urbanost« v primeru Ljubljane in Maribora. Teorija in praksa, 52
Uršič M, Hočevar M (2007) Protiurbanost kot način življenja. Fakulteta za družbene vede
Uršič M, Tiran J (2022) Vpliv prostora na volilno vedenje: analiza delitve med urbanim in ruralnim na primeru volitev v Državni zbor 1996–2022. Teorija in Praksa 59(4):871–1024
Weiss RS (1995) Learning from strangers: the art and method of qualitative interview studies. Simon and Schuster
Weston K (1995) Get thee to a big city: sexual imaginary and the great gay migration. GLQ: A J Lesbian Gay Stud 2(3):253–277
Wimark T, Lewis NM, Caretta MA (2017) A life course approach to the field and fieldwork. Area 49(4):390–393

Correction to: How Can Scientific Literacy Be Enhanced in Geography Education? Two Ways Forward: Geoboxes and Digital Experiments

Nina Scholten, Natalie Bienert, and Rainer Mehren

Correction to:
Chapter 5 in: D. Piróg and A. Świętek (eds.), *Future-Ready Geography*, Key Challenges in Geography, https://doi.org/10.1007/978-3-031-81268-2_5

In the original version of Chapter 5, the caption of Figure 1 has been updated and a new reference has been added. The chapter has been updated with the changes.

The updated version of this chapter can be found at
https://doi.org/10.1007/978-3-031-81268-2_5

© The Author(s), under exclusive license to Springer Nature Switzerland AG 2025

D. Piróg and A. Świętek (eds.), *Future-Ready Geography*, Key Challenges in Geography, https://doi.org/10.1007/978-3-031-81268-2_13

Author Index

A
Aziz, Asad, 177

B
Baarová, Barbara, 153
Bargieł, Kinga, 101
Bienert, Natalie, 85
Brede, Lara, 133

C
Cichoń, Małgorzata, 153
Cybul, Piotr, 41

E
Eberth, Andreas, 133
Enikiev, Dimitar, 215

H
Hawke, Shé Mackenzie, 237

I
Ilieva, Nadezhda, 215

K
Kazakov, Boris, 215
Kilar, Wioletta, 23
Klonari, Aikaterini, 65
Kumer, Peter, 237

M
Mehren, Rainer, 85
Meyer, Christiane, 133
Montebelli, Stefania, 195

P
Piotrowska, Iwona, 3

S
Scholten, Nina, 85
Świętek, Agnieszka, 41

T
Tsvetkov, Kaloyan, 215

Z
Zisi, Christina, 65

© The Editor(s) (if applicable) and The Author(s), under exclusive license to Springer Nature Switzerland AG 2025
D. Piróg and A. Świętek (eds.), *Future-Ready Geography*, Key Challenges in Geography, https://doi.org/10.1007/978-3-031-81268-2

Subject Index

C

Change agents, 133, 134, 136–138, 141–149
Competence of the future, 166
Competences, 15, 23, 24, 30, 31, 33–37, 46, 58, 86, 146, 153–159, 161–172
Containment of land consumption, 195
Critical education events, 101

D

Development, 4, 7–11, 13, 15–18, 23–26, 30, 31, 33, 35–37, 41–43, 47, 50, 52, 67, 69, 80, 88, 90, 91, 104, 105, 108, 111, 135, 139, 153, 154, 156, 157, 159, 166, 168–171, 177, 178, 180, 182–184, 186, 195, 196, 198, 200–204, 207–211, 216, 218, 219, 228, 232, 240, 241
Discussion group, 101–109, 113, 126, 128

E

Educating city, 198–200
Education about the territory and active citizenship, 200
Education for Sustainable Development (ESD), 15, 17, 133, 134, 136–140, 148, 149, 155, 156, 170, 171
Education in Poland, 24, 32, 121
Experiments, 50, 85, 86, 88–98, 184

F

Facebook, 101, 105, 108, 109

G

Generation alpha, 41, 53
Generation Z, 41, 49, 52, 105
Geographical education, 3, 13, 23, 24, 31, 32, 34–36, 41, 56, 112
Geographical space, 3, 32
Geographic lifelong learning, 31, 195
Geography of heritage, 11, 41, 44
Geography of thinking, 15
Ghettoization, 216, 217
GLBTQI+, 238

H

Heritage, 11, 41–48, 50–58, 202, 209, 248
Heritage education, 41, 43, 44, 46–49, 52, 56–58

I

Inquiry based learning, 85

K

Kindergarten pupils, 65

L

Labour market, 23, 24, 26–37, 233
Large-scale giant maps, 68–70, 78
Learning conditions, 153, 154, 159, 160, 162–165, 171

M

Metronormativity, 238, 239

© The Editor(s) (if applicable) and The Author(s), under exclusive license to Springer Nature Switzerland AG 2025

D. Piróg and A. Świętek (eds.), *Future-Ready Geography*, Key Challenges in Geography, https://doi.org/10.1007/978-3-031-81268-2

P
Population, 8, 105, 177–181, 183–185, 201, 209, 215–218, 221–228, 233, 239–241, 243–246, 250, 251
Post-growth, 133–136, 138–140, 142, 143, 146–149

R
Rural in-migration, 237–239, 250
Rural subjectivities, 239

S
Science kits, 85
Scientific inquiry, 9, 18, 86, 88–90, 92–94, 97, 98
Scientific literacy, 85–88, 90–94, 96–98
Sexuality and space, 244
Shared urban regeneration, 195
Social media, 49, 50, 56, 101–108, 111
Social segregation, 215, 216
Socio-economic geography., 10, 14, 17, 23, 24, 36
Spatial abilities, 65–69, 77–80
Spatial polarization, 215, 216
Sustainability, 16, 17, 139, 145–149, 154, 156, 158, 162, 165, 166, 171, 177–187, 196
Sustainable development, 8, 14–17, 43, 134, 135, 137, 153–157, 159–163, 165–172, 180–187, 196, 202, 207

T
Teachers of geography, 42, 101, 109
Teaching interventions, 65, 68, 69, 72, 74, 75, 77–80
Transformative literacy, 133, 135, 137

U
University, 3, 9–12, 17, 29, 56, 67, 69, 94, 101, 105, 107, 133, 153, 154, 156–160, 162, 165, 166, 168, 169, 171, 172, 200, 205, 207, 244
Urban, 26, 48, 138, 143, 177, 179, 181–187, 195–207, 209–211, 215–218, 220–225, 229, 233, 234, 237–245, 248–251
Urbanization, 177–181, 183, 184, 187, 210, 216, 240, 241
Urban Roma, 206, 216

GPSR Compliance

The European Union's (EU) General Product Safety Regulation (GPSR) is a set of rules that requires consumer products to be safe and our obligations to ensure this.

If you have any concerns about our products, you can contact us on ProductSafety@springernature.com

In case Publisher is established outside the EU, the EU authorized representative is:

Springer Nature Customer Service Center GmbH
Europaplatz 3
69115 Heidelberg, Germany

Batch number: 10755227

Printed by Printforce, the Netherlands